THE HEAVY WATER WAR

THE HEAVY WATER WAR

BEATING HITLER TO THE BOMB

JOHN SADLER

AMBERLEY

This is for Bev.

First published 2024

Amberley Publishing
The Hill, Stroud
Gloucestershire, GL5 4EP

www.amberley-books.com

ISBN 978 1 3981 1767 9 (hardback)
ISBN 978 1 3981 1768 6 (ebook)

British Library Cataloguing in Publication Data.
A catalogue record for this book is available from the British Library.

1 2 3 4 5 6 7 8 9 10

Typesetting by SJmagic DESIGN SERVICES, India.
Printed in the UK.

Since the completion of the first atomic bomb nothing has been accomplished to make the world safer from war, while much has been done to increase the destructiveness of war. I am not able to speak from any first-hand knowledge about the development of the atomic bomb since I do not work in this field. But enough has been said by those who do to indicate that the bomb has been made more effective. Certainly, the possibility can be envisaged of building a bomb of far greater size, capable of producing destruction over a larger area. It also is credible that an extensive use could be made of radioactivated gases which would spread over a wide region, causing heavy loss of life without damage to buildings.

But it is not necessary to imagine the earth being destroyed like a nova by a stellar explosion to understand vividly the growing scope of atomic war and to recognize that unless another war is prevented it is likely to bring destruction on a scale never before held possible and even now hardly conceived, and that little civilization would survive it.

Albert Einstein

The heat was tremendous. And I felt like my body was burning all over. For my burning body the cold water of the river was as precious as the treasure. Then I left the river, and I walked along the railroad tracks in the direction of my home. On the way, I ran into another friend of mine, Tokujiro Hatta. I wondered why the soles of his feet were badly burnt. It was unthinkable to get burned there. But it was undeniable fact the soles were peeling, and red muscle was exposed.

Even I myself was terribly burnt; I could not go home ignoring him. I made him crawl using his arms and knees. Next, I made him stand on his heels and I supported him. We walked heading toward my home repeating the two methods. When we were resting because we were so exhausted, I found my grandfather's brother and his wife, in other words, great uncle and great aunt, coming toward us. That was quite a coincidence. As you know, we have a proverb about meeting Buddha in Hell. My encounter with my relatives at that time was just like that. They seem to be the Buddha to me wandering in the living hell.

Akihiro Takahashi. He was 14 years old when the bomb was dropped on Hiroshima.

CONTENTS

PREFACE

Went the day well?
We died and never knew.
But well or ill,
Freedom, we died for you.

John Maxwell Edmonds

If Hitler had succeeded in developing a nuclear bomb, that could have been both the end of the Second World War and of human civilisation as we know it. A handful of SOE-trained Norwegian Commandos stopped him.

During the course of the War, the Allies mounted a series of attempts to halt or prevent Germany from manufacturing heavy water utilising hydroelectric plants in occupied Norway. These efforts comprised of a mix of bomber and Commando raids. The overall aim was to stop Nazi Germany building a nuclear bomb. In fact, Hitler was never as close to success as the Allies thought, but the idea that his regime could construct and deploy such a device was the ultimate Domesday scenario. The mere threat would have been sufficient to force a negotiated peace. A Nazi bomb hanging over the world like a nuclear sword of Damocles would have been the worst threat to humanity between the Ice Age and the Cold War.

Production, and therefore the Allied target, centred on the Vemork Power Station standing by the Rjukan Waterfall at Telemark, an upland industrial region west of Oslo. The hydroelectric plant, commissioned in 1934, was the first in the world capable of the mass-manufacture of heavy water, with a capacity of 12 tonnes per annum. Just before the Germans successfully invaded Norway in April 1940, the French Secret Service, *Deuxième Bureau*, spirited away 408 lb (185 kilograms) of heavy water. Norway was officially neutral, but the plant's CEO agreed to 'lend' France the product for the duration of hostilities. It was moved first to Oslo, then via Perth to France. This removed any immediate threat, but the plant still retained its full production capacity. As the Fall of France became imminent the heavy water was successfully shipped to England, a significant and dramatic exploit in itself.

In early 1943, a series of daring raids, Operations *Grouse*, *Freshman, Swallow* and *Gunnerside*, neutralised the plant's capacity. During October 1942, in the first of these missions, Britain's SOE dropped a team of four Norwegian operatives onto the inhospitable Hardanger Plateau, above the plant itself. Next month, the second phase, *Freshman* was launched; an attempt to land paratroops on the plateau who would then link up with the Norwegian quartet to attack Vemork. This proved disastrous, gliders bringing in the paras crashed along with one of their Halifax tugs. Casualties from the crashes, plus those captured, tortured and killed by the Germans accounted for 100 per cent of the glider-borne force. By this time, the notorious *Kommandobefehl* was already in place. Germany refused to treat Commandos, despite their being in uniform, as legitimate combatants but rather as partisans or 'terrorists' – simply to be shot out of hand.

In February 1943 a crack force of SOE-trained Norwegian Commandos succeeded in sabotaging the plant's production capacity. Further manufacturing effort was abandoned, and the Nazis attempted to transport the heavy water they had aboard the ferry SF *Hydro*. The Norwegians managed to sink the vessel in the deep waters of Lake Tinn (despite best efforts there were civilian casualties) and Germany's chances of constructing a nuclear device effectively went down with the ship.

The Domesday threat was removed.

In the November 2004 Remembrance ceremony, I had the privilege of marching from the village at Spean Bridge up to the Commando Memorial – I did so in honour of a brave Dane, Jack Rasmussen, who had served with No. 4 Commando throughout the war. It was a blustery autumn day, typical of the Western Highlands at that time of year, dreich autumn with winter just held at bay. This would be the last occasion when veterans marched together. And march they did, swinging up the steep climb from the settlement to the marvellous panorama where their comrades, immortalised in bronze, look out over a sublime sweep of Scottish landscape. They were magnificent, these men in their eighties. If one imagines them in their twenties, no wonder Hitler was afraid of them!

M.R.D. Foot, Special Operations Executive ('SOE') historian, says this of Operation *Gunnerside*: 'One coup SOE brought off on Norway changed the course of the war; and was so important that nothing else the section did bears record beside it.' He goes on to describe the course of Operations *Grouse/Swallow*, *Freshman* and *Gunnerside*, concluding: 'If SOE had never done anything else, *Gunnerside* would have given it claim enough on the gratitude of humanity.'[1]

Major (now Major-General) Frode Kristoffersen of the Norwegian Armed Services stressed in his 2010 paper the relevance these operations have to modern Special Forces doctrine:

> As another example of the effectiveness of special operations over more conventional approaches, it is important to consider subsequent operations at Vemork. German efforts to reconstruct the heavy water plant in the nine months following Operation *GUNNERSIDE* led the Allies to launch a new conventional operation: the bombing of Vemork in November 1943. The result of the bombing proved true the challenges of the operating environment at Vemork as the underground heavy water installations remained intact. Additionally, the bombing resulted in collateral damage, including twenty Norwegian civilians killed at Rjukan.

> Again, SOE had more success in a following mission carried out by SOF personnel. The sinking of a Norwegian ferry carrying heavy water caused civilian casualties but achieved the objective of denying Hitler this strategic resource. Patrol sized operations achieved a strategic outcome with their efforts in February 1943, with a precise hit on the most vulnerable point of the heavy water production that SOE accomplished without losing personnel and without causing collateral damage. The small group of selected SOE soldiers capitalized on their specialized training, skills, and their unique motivation, when carrying out their mission in an extremely difficult operating environment.'
>
> The planning and execution of Operations *GUNNERSIDE* and *GROUSE* were textbook SOF operations that illustrate today' s theories of special operations. Like SOE during World War II, today's SOF units should carefully select, train, and employ only the limited number of soldiers that are able to fully master the operating environment. By mastering the operating environment and embracing the sources of friction rather than accepting their restrictions, SOF units can contribute to the current fight as flexible and creative elite warriors.[2]

I have extended family in Norway and have made the journey there many times by both sea and air. It is a remarkable country, jagged fjords of breathtaking beauty but at the same time hard and uncompromising, nowhere more so than the Hardanger Plateau, which features so extensively in the story of the Telemark raiders. Merely to go there offers an insight into what these very remarkable men took on, endured and survived.

ACKNOWLEDGEMENTS

Thanks are due to all of the following without whose generous assistance this book could not have been written: Catriona Fleming of Clan Cameron Museum; Ivar Kraglund & Frode Faeroy of the Norwegian Resistance Museum, Oslo; Vanessa Martin of the West Highland Museum, Fort William; Major S. T. Meadows of 2nd Battalion Royal Gurkha Rifles; Claire Grint of Cogito Books, Hexham; staff at the Lit & Phil Museum, Newcastle upon Tyne; Bob Brooks of Hotspur School of Defence, Alnwick; Rebecca Blackburn of the Royal Engineers Museum, Chatham; Greethe Flood at the Norwegian Defence Historical Archive, Lars Ohrstrom of Chalmers University of Technology, Major-General Eirik Johan Kristoffersen of the Norwegian Armed Forces, The Royal Marines Charity, the US Central Intelligence Archives for consent to quote from the feature by Claus Helberg, staff at the National Archives, Kew and at the National Army Museum, Chelsea.

My thanks are due also to my agent Tom Cull, to Shaun Barrington and Nicola Embery at Amberley, to Julia Grint for editing and to my remarkable research assistant Bev Palin for modelling, research, typing and indexing and to whom this book is dedicated, to Paul Macdonald on Fairbairn-Sykes fighting knives and of course and as ever, to my wife Ruth for enduring the genesis of yet another book.

I'm obliged to Professor Lars Ohrstrom of Chalmers University of technology for permission to quote from his feature 'Leif

Tronstad – the Heavy Water Hero' (*Chemistry World*, February 2018) and to Major Frode Kristoffersen for permission to quote from his Master of Military Studies dissertation, *Operations Gunnerside and Grouse – Special Operations During World War Two against the German Controlled Heavy Water plant in Norway* (USMC Command Staff College, April 2010). The testimony of James Robert Ward is recorded as part of the Westall Project; http://www.northshields173.org, (retrieved 8 May 2023). *Air Raid* is reproduced by courtesy of The Northumberland Hussars and 15/19 Hussars Archive (TWAM). 'Epitaph on a New Army' is from Selwyn, V., *More Poems of the Second World War* (London, Everyman 1989), '*Taint* (extract) appears by kind permission of the Trustees of the Fusiliers Museum of Northumberland, Kipling's *Edgehill Fight* features in Fletcher, C.R.L., *History of England* (Oxford, Clarendon Press 1911), *These are Your Sons* appears by kind permission of Durham County Record Office. Information on the controversial Sark Raid of October 1942 is taken from Warwicker, J., *Churchill's Underground Army* (London, Frontline 2008), chapter 21. 'After Party' is by kind permission of *Nuclear war Poems - Modern Award-winning Nuclear war Poetry.*

I have striven to provided proper acknowledgement and accreditation wherever possible. Any failure to do so is an error on my part and should anyone notify me of such mistakes, I will ensure due correction is made in any and all further editions.

John Sadler
Northumberland

LIST OF PHOTOGRAPHS AND ILLUSTRATIONS

8. Queues for food in Oslo, 1942. Shortages caused suffering among the general populace. (Library of Congress)
9. Norwegian refugees make their way across the porous border with Sweden, escaping the occupation. (Arkivverket)
10. Young Norwegian men in Oreyd refugee camp, Sweden. (Arkivverket)
11. The imposing Vemork power plant outside Rjukan, Telemark. Vemork was the first plant in the world to mass-produce heavy water. (National Library of Norway)
12. Svelgfoss power plant in Notodden, Telemark. (Norwegian Folk Museum)
13. Saheim power plant (or Rjukan II) in Rjukan, Telemark. (Norwegian Folk Museum)
14. Pipelines at Rjukan demonstrate the scale of the operation. (Norwegian Folk Museum)
15. Leif Tronstad in 1940, probably photographed after the invasion. By this time Tronstad had already organised a line of defence with volunteer marksmen covering the retreat of Norwegian forces at Østerdalen, and he was rapidly involved in resistance efforts. (National Library of Norway)
16. Kompani Linge in 1945. Back, left to right: Oddvar Dobloug, Erling Jensen and Joachim Rønneberg. Front, left to right: Martin Olsen, Birger Rasmussen, Ragnar Ulstein and William Johannesen. (Musea i Nord-Østerdalen)
17. Tronstad (bottom left) along with, left to right, King Haakon VII, Harald Sandvik and Crown Prince Olav at Glenmore Lodge, Scotland, where Kompani Linge were trained. (Henrik Aamodt)
18. The site of the Fieldfare cabin used by the saboteurs. The original cabin can be seen in the picture upon patient examination. (Frankemann)
19. SF *Hydro*, the steam-powered railway ferry used to transport materials from Rjukan to the port at Skien. The saboteurs sank it in the deepest part of Lake Tinn, dealing a massive blow to Nazi efforts to stockpile heavy water. (Norway National Library)

MAP

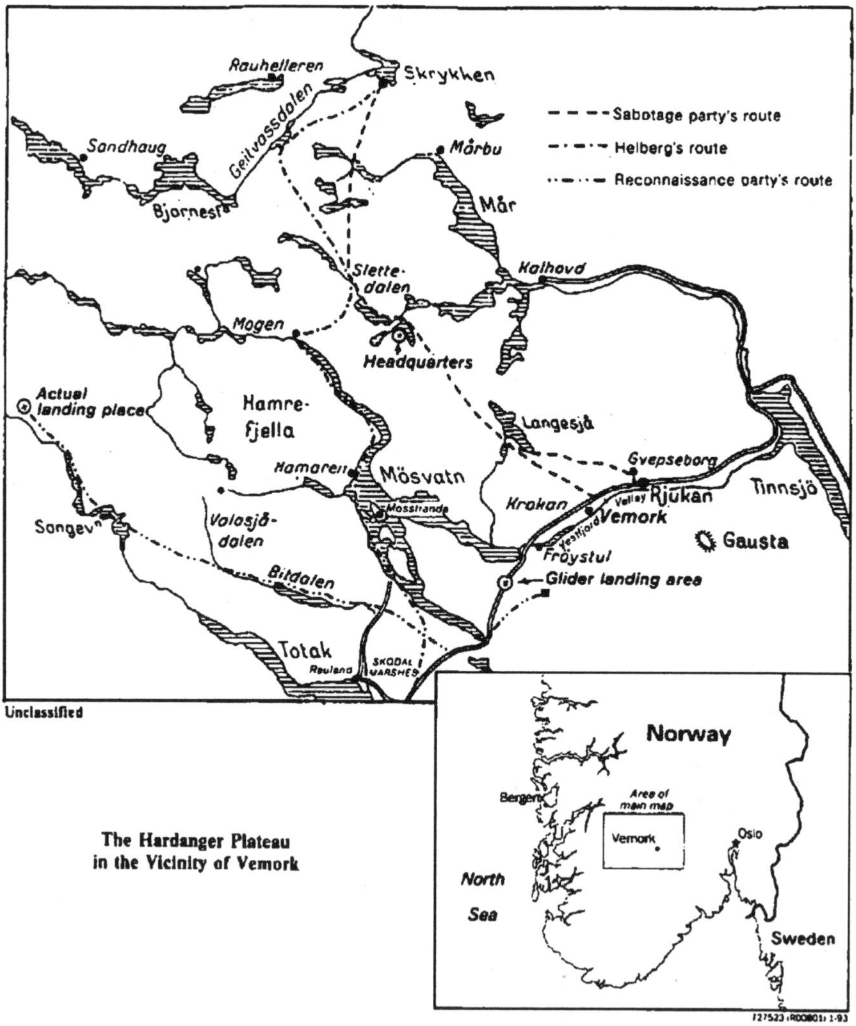

The Hardanger Plateau
in the Vicinity of Vemork

DRAMATIS PERSONAE

Norway

Professor Leif Hans Larsen Tronstad (1903–1945) DSO OBE, was one of five children, largely raised by his mother. By 1918, when he graduated from school, his talents were clear. He became a student at what would become Oslo University's Faculty of Engineering. Having switched to technical chemistry, he graduated first in his year.

In 1924, he moved up to Trondheim to study. He completed his compulsory military service and was promoted to 2nd Lieutenant armourer. He was appointed Professor of technical inorganic chemistry at the Norwegian Institute of Technology in 1936.

Tronstad worked partly as a consultant for Norsk Hydro. When the Germans invaded, he was, of course, immediately called up. When the fighting was over and with Norway occupied, he joined the fledgling resistance movement. Tronstad began sending London information on the Germans' growing interest in the plant. In September 1941 he escaped firstly to Sweden and then London.

He first joined the Ministry of Defence and latterly the High Command of the Norwegian Government in Exile. From 1943, he headed up a section of 4 (FOIV), set up a year earlier under the leadership of Colonel Bjarne Oen. This brought a step up from three pips to a crown. His brief was wide-ranging, from operations involving industry and shipping, training the Norwegian Independent Commando Company (*Kompani Linge*), advice on sabotage and, towards the end of hostilities, further advice on safeguarding Norwegian industries from destruction by the Germans as they withdrew. As part of his expansive role, he set up the Norwegian High Command Technical Committee, bringing together the formidable talents of other exiled scientists He did finally get his wish to go an active service but with tragic consequences (see chapter ten).

Martin Jensen Linge (1894–1941) DSM, was a native of Norddal. He trained at Trondheim Military Academy and came out as an NCO, later commissioned as lieutenant in the reserves. He went on to act and had a successful stage career during the inter-war years. Recalled to the colours in 1940, he was nominated as liaison officer with the British on Andalsnes Island. It was through there that the King, cabinet, and senior command escaped to England with the national gold reserve. So did Linge, though he was in fact casevaced after being wounded, the first Norwegian military casualty to reach England.

Once recovered, he teamed up with other escapees and lobbied for the creation of a functioning resistance, active in raiding and sabotage – he argued that the nature of the terrain made his country a natural bridgehead. By August 1940, he was the official liaison with the War Office and recruiting manpower for what became Norwegian Independent Company no. 1. Linge was no armchair warrior as he had already demonstrated, and despite his age, he insisted on going on ops. He lost his life during Operation *Archery*. In his memory, his unit was renamed *Kompani Linge.*

Einar Skinnarland (1918–2002) DCM was a native of Telemark and a graduate from the Engineering College in Porsegrunn. He worked at Vemork (as did his brothers and numerous friends) but escaped to the UK aboard the 'requisitioned' (hijacked) steamer *Galtesund* in 1942. He immediately joined *Kompani Linge*. He was the first SOE agent to be infiltrated into Rjukan, parachuted down onto the the Hardanger Plateau in March 1942.

It was Einar Skinnarland who, during his preliminary de-briefing, convinced his SOE handlers that bombing Vemork simply wouldn't work and that only a surgical raid could complete the job. He made a major contribution to the successes of *Grouse/Swallow* and *Gunnerside*. He saw out the war in his native land, leading the resistance in Rjukan and Telemark.

After 1945, he was garlanded with a slew of honours and subsequently moved to the US to spend the next 40 years as a civil engineer working on international projects.

2nd Lieutenant Jens-Anton Poulsson (1918–2010) DSO, a native of Tinn in Telemark, had just joined the Norwegian Army in time for the invasion but escaped to Britain. This proved an adventure in itself. He got clear across into Sweden but there was no direct air connection to the UK, so he tabbed through Finland, Soviet Russia and as far south as Turkey, a journey worthy of his Viking ancestors. He still wasn't done and his passage to Britain took him through Syria, Lebanon, Palestine and Egypt, where he took ship to India and from there to South Africa. Next, he journeyed across the oceans to reach Trinidad, thence to Canada and finally, via the regular convoy system to Liverpool. It took him nine months. An experienced mountaineer of impressive physique with his trademark pipe permanently clamped in his teeth, he was the natural choice to lead *Grouse*.

Aside from his distinguished role as leader of *Grouse/Swallow*, he went back again in 1944, alongside Tronstad as part of Operation *Sunshine*. The brief was to set up fighting contingents of resisters at Rjukan, Notorden and Nore. He raised and equipped a contingent of 300 before taking over as leader of the mission on Tronstad's death. His successes prompted a series of awards.

After the war he remained in the Norwegian military and commanded a joint Danish Battalion (DANOR) during peacekeeping operations in Gaza. From 1961 until his retirement, he held a series of senior positions, even commanding the King's personal Guard 1961–1965.

Knut Magne Haughland (1917–2007) DSO MM, was a native of Rjukan who just before the war studied military radio transmissions and joined the Norwegian Army. He subsequently fought in the battles for Narvik and after the fall, while doing factory work in Oslo, joined the swelling resistance movement. In 1941, he was arrested but made good his escape, getting to Britain through Sweden. He became a natural recruit for *Kompani Linge*.

For a period, after the Vemork raid, he stayed on in Norway, working with Einar Skinnarland whom he trained up as a radio operator. After a brief return to Britain for supplies, he was again parachuted back into Norway in November 1943. This was as risky as it gets. He was nabbed a second time but again managed to get away. His luck continued to hold, even after one of his transmitters was pinpointed by radio direction finding. He ran just in time and returned safely to the UK.

After the war he was awarded Norway's highest decoration, the War Cross with Sword and effectively with bar, earning the medal in 1943 and again in 1944. In addition to his British decorations, he picked up a *Croix de Guerre* and other medals. He spent the rest of his career in the Armed Forces, transferring to the Air Force, retiring in 1963 to take up a post as the first curator of Norway's Resistance Museum. In 1947, as if he'd not had enough excitement during the war, he joined Thor Heyerdahl and the Kon-Tiki Expedition.

Claus Helberg (1919–2003), another native of Rjukan, was born to the outdoor life and by 1938 was working locally as a tourist guide. Captured in 1940 while serving, he managed to escape and arrived in Britain early in 1942. He was recruited into *Kompani Linge*, a natural candidate for the *Grouse* team. He certainly didn't disappoint and if he had not existed Hollywood would have to have invented him, the *D'Artagnan* of saboteurs, adept at pushing his luck to the extreme and managing to get away with it. He is the basis, albeit loosely, for Richard Harris's character in *Heroes of Telemark*.

Arne Kjelstrup (1913–1995) MM, another native of the Rjukan area and a plumber in his pre-war civvy job was called up in 1940. He stopped a Mauser round early on but survived as a pair of wire cutters he was carrying absorbed much of the shock of the bullet (though it remained lodged in him for the rest of his life). He escaped after the surrender and, like Poulsson, experienced a long, circuitous trip to reach Britain, another recruit for *Kompani Linge*.

After the Vemork raid, he stayed on, building up resistance cells working with *Milorg* (*Militaer Orginasjon* ('Military Organisation'), the official Norwegian Resistance (see chapter three). In Western

Telemark it was the *Bonzo* Group, named after 'Bonzo' Haukelid, renowned for his fearlessness and who had a teddy bear of that name as a boy!

He subsequently returned to the UK, honing his skills as a saboteur. He joined Tronstad for Operation *Sunshine*. His last wartime mission was Operation *Starlight* raising an active force of nearly a thousand armed volunteers to harass the wavering occupiers. During this operation, he met his future wife Tulla Irgens and, once the war ended returned to his civvy job in Baerum, dying just a few days before the 50th anniversary of the liberation he had done so much to win.

Lieutenant Joachim Holmboe Ronneberg (1919–2018) DSO, was born in Alesund to a well-connected family. Called up as a military surveyor in 1938, he was mobilised when the Germans invaded. In 1941, with the Axis firmly in control of the country, he, together with a party of eight others, escaped by boat to Scotland, another willing recruit for *Kompani Linge*. His leadership skills, grit and determination soon marked him out and earned him his commission. From the start, he commanded the *Gunnerside* team.

Following the successful strike against the Vemork factory, he won the War Cross with Sword, was awarded the St. Olav's medal with Oak Branch, the *Legion d'honneur*, the *Croix de Guerre* plus the US Medal of Freedom with Silver Palm. This was by no means the finish of his wartime career. He commanded a further series of surgical raids including Operation *Fieldfare* – this took place between early March and May 1944. In January, he led a strike aimed at blowing up the Stuguflat railway bridge. He succeeded, knocking the crossing out for several weeks. Once again, his team made good their escape without incurring any casualties.

After the war, he worked extensively in the media and presented regular lectures on the war years. In 2014, four years before his death at the age of 99, a bronze statue of him by sculptor Hakon Anton Fageras was unveiled in his home town of Alesund by Princess Astrid.

Knut Haukelid (1911–1994) DSO MC, was something of the odd man out as he was born in Brooklyn – his father was a civil engineer. They moved back to Oslo in 1914 where Haukelid snr worked on the Metro. A further stint in the USA followed. Knut attended Massachusetts Stare College, completing his engineering studies at Dresden School of Technology and the University of Berlin. Escaping after the Germans invaded, he was another who found his way into the ranks of *Kompani Linge*, acting as Ronneberg's 2/IC.

After the success of *Gunnerside*, he struck the second, massive blow to the Nazi programme by sinking the ferry transporting reserves of heavy water SF *Hydro* on 20 February 1944. Once Norway was free, he rejoined the army and stayed on in various senior posts until his retirement in 1974, by which time he held the rank of Lieutenant-Colonel, Knut Haukelid died in 1994 at the age of 83.

Kaspar Idland (1918–1968) MM, was born in the village of Figgio (now a suburb of Sandnes in Rogaland, Western Norway), one of eight children. He joined the Norwegian Army early on and earned his sergeant's stripes. He fought through several of the 1940 battles and was for a while held as a PoW. Released in September 1941, he escaped by boat to Scotland and joined *Kompani Linge*. He saw action with them during raids on Reine and Moskenes (Operation *Anklet*), December 1941.

Following success at Vemork he returned to Norway during 1944–1945, part of resistance group Vestige 4. As the war drew to a close, from March 1945 Idland commanded one of the Varg network groups, based in a mountain *saeter*, *Langelona*, which was still growing when hostilities finally ceased. He lived in Stavanger for a decade before emigrating to the US, residing in Huntingdon, New York. He, too, has a memorial in his home town of Figgio (Sandnes).

Lieutenant Fredrik Thorbjorn Kayser (1918–2009) MM, was from Paradis near Bergen on the west coast. An early recruit for the Boy Scouts, his team leader was the later resistance hero Fredrik Rieber-Mohn. Called up in 1938, he served as a member of the

King's Guard. He got his first taste of a shooting war fighting with the Finns against would-be Russian Communist oppressors. He never in fact fired a shot but returned to take up arms against the Nazis, and this time he had plenty opportunities to use his rifle, taking part in several actions.

Beaten but undefeated, he became involved in the early stages of resistance in his native Bergen, helping to run the underground press. He soon tired of this and craved a more active role – escaping to the UK in September 1941. Like his future comrades he joined Linge, honing his lethal skills at SOE finishing school. He gained operational experience on a couple of abortive raids but was clearly prime material for *Gunnerside*. He got back to Britain following the Vemork raid to be retrained in the use of one-man midget subs.

During the final year of the war when it was clear Norway would, at some point, be liberated, SOE planned to create a series of safe harbours in suitably remote locations as refuges for resisters and those avoiding forced labour. Kayser at this point having earned his commission, found himself once again in a battle for the *Bjorn West* bastion at Masfjordan in Central Vestland. The fight lasted for nearly a week as the resisters withstood German efforts to dislodge them. All in all, a very active war.

Kayser received no fewer than fourteen decorations for his service, including the St. Olav's Medal with Oak Branch and bar. Sadly, he didn't live quite long enough to witness the unveiling of his memorial at Storetveit in Bergen in May 2012.

Hans 'Kyllingen' Storhaug (1915–1995) MM, DSM, was born in Rena near Asta, Innlandet County. He fought against the Germans in 1940 and, like so many others, refused to submit to the jackboot so escaped to Scotland, joined Linge in time to participate in Operation *Anklet*, a raid on Reine (Lofoten Islands) in December 1940; another natural for Team *Gunnerside*. He escaped through Sweden but was back in action in October '43 during Operation *Grebe*. Three out of six team members drowned when they parachuted into a frozen lake. Storhaug with the other survivors made it out, once again via Sweden, infiltrating back into Norway in March 1944 for Operation *Grebe Red*, and fighting behind enemy lines until war's end.

Birger Edvin Martin Stromsheim (1911–2012), was a native of Alesund, a builder by trade. In 1941 he and his wife both escaped to Scotland courtesy of the 'Shetland Bus'. He joined *Kompani Linge* and was selected for *Gunnerside*. Ronneberg reckoned he was one of the most steadfast recruits. In such company, that is high praise indeed.

After the Vemork Raid and his successful escape through Sweden, he was awarded the St. Olav's Medal with Oak Branch, presented to him by King Haakon personally at STS 26 near Nethy Bridge, Speyside. He made it past his century dying aged 101 in Oslo, where he is buried in Nordstrand Church.

Knut Lier-Hansen (1916–2008) was another native of Rjukan and served as an NCO in the Norwegian army during the German invasion, where he saw action at Gransherad. An early member of the underground, his fame rests on his role in the sinking of SF *Hydro*. Despite the fact the Norwegian Government in Exile had approved the ferry sinking and its importance in the fight against Germany, the loss of civilian lives rankled, and he continued to receive death threats on the anniversary for many years after.

Odd Kjell Starheim (1916–1943) DSO was the son of a ship's captain and served pre-war for six years in the Norwegian Merchant Navy. He fought on land in 1940, was captured and escaped. Subsequently, he managed to get out of Norway by a daring sea passage to Aberdeen, where he quickly joined Martin Linge. For SOE, he undertook a series of missions back on home turf. This was inevitably perilous and led to his arrest in Oslo by the Gestapo. That might easily have been the end, but he leapt to freedom through a second-storey window.

Norway was now too hot which led, in conjunction with Einar Skinnarland, to the audacious hijack of *SS Galtesund* and their successful dash back to Aberdeen. In February 1943, he led a larger and more ambitious raid, Operation *Carhampton*, which ran into stiff opposition. He with a dozen surviving Commandos attempted to repeat his earlier feat by seizing SS *Tromosund.* Third time wasn't lucky, and the ship was sunk with all hands. His was one of only two bodies recovered and he is buried in his native Lista. As well as the DSO he was awarded the Norwegian War Cross.

Britain

Major-General Colin McVean Gubbins (1896–1976) KCMG DSO MC, had been commissioned into the RA in 1914, but with the outbreak of war imminent was studying German at Heidelburg. Bad timing. To avoid internment, he dashed back in true John Buchan style via Belgium, to reach England on 3 August. He had a very active war, seeing a great deal of combat and being wounded and gassed several times. After the war he served in North Russia, then during that Hedgerow War in Ireland against the IRA. At the end of the conflict, it was he who gave artillery to Michael Collins to blast dissidents at the start of the Irish Civil War. He also (reluctantly) provided the gun carriage for Collins' funeral.

Hugh Dalton (right), Minister of Economic Warfare, and Colin Gubbins, chief of the SOE, talking to a Czech officer during a visit to Czech troops in Warwickshire. (Crown copyright)

Gubbins found he was developing a taste for this type of highly irregular conflict. He held a variety of staff appointments between the wars and in 1939 served under the legendary Adrian de Carton Wiart, leading the military mission to Poland. This didn't last long and in March 1940 he was summoned back from Paris to home service raising those independent companies who would become the Commandos. He saw active service in the abortive Norwegian campaign but was soon back in England organising the stay-behind Home Guard special forces.

It was Hugh Dalton, Minister of Economic Warfare who wanted Gubbins for the newly formed SOE. He had three specific tasks: (1) establish suitable training sites, (2) coordinate operational procedures and (3) liaise with the Joint Planning Staff. His dogged persistence and operational genius allowed SOE to flourish and begin to succeed, despite much opposition within conventional branches of the services. The Vemork Raid would be the most stellar of these triumphs.

Lieutenant-Colonel John ('Jack') Skinner Wilson (1888–1969), a Scot, was one of those pre-war advocates for the scouting movement and a friend of Baden-Powell (he'd been acting director from 1938–1939). He was a pupil at Glenalmond before joining the Colonial Police Service in the sub-continent, developing a specialist role in counter-insurgency operations. He also had something of a reputation as an imperial dragon- or rather leopard-slayer, *Baghmara*.

His primary role within SOE was training, but from 1942 he headed up the Norway section. M.R.D. Foot characterised him as having the patience, accuracy and calm of a good scoutmaster! In recognition of his wartime achievement Haakon VII awarded him the Royal Norwegian Order of St. Olav. After the war he returned to scouting and was very active in the movement right up to his retirement.

Germany

General Nikolaus von Falkenhorst (1885–1968) had joined the army in 1903 serving in a variety of roles throughout the Great

War, followed by stints in the paramilitary *Freikorps* and then the *Reichswehr*. During the 1939 invasion of Poland, he commanded XXI Corps. On 20 February 1940, the Fuhrer appointed him to lead the invasion of Norway.

He suffered losses to his ships during the naval battle at Narvik, but a limited and rather costly Allied victory could not affect the outcome. Success gained him further promotion (to the rank of *Generaloberst*). In December 1944 when he was relieved and relegated to the strategic reserve. After the war, he was tried for alleged war crimes by a joint Anglo-Norwegian tribunal, convicted and sentenced to hang. This was subsequently commuted to 20 years gaol time, but he only served seven before being released.

***Reichkommissar* Josef Terboven** (1898–1945), born in Essen into a minor gentry family, was an early convert and joined the Nazi Party in 1923, in time to take part in the debacle of the Munich *Putsch*. He kept his head down for a while but re-joined in 1925 and served in Rohm's SA, rising to local prominence and the provincial status of *gauleiter*.

He was appointed to his office in Norway on 24 April 1940, He controlled security forces numbering perhaps 6.000. On 8 May 1945, with the Reich in ruins he committed suicide by detonating an explosive charge inside his last-ditch bunker.

General Wilhelm Rediess (1900–1945) was the local Higher SS and Police commander in Norway throughout the occupation. Born the son of an imperial court functionary, he became an electrician, serving briefly in the latter stages of the Great War as an infantryman. At the outset of war, he was in charge of deporting all Jews from East Prussia. In Norway, he served alongside Terboven and committed suicide in the same bunker.

***SS Oberfuhrer* Colonel Heinrich Fehlis** (1906–1945), a native of Gottingen, trained as a lawyer. He joined the SA in 1933 as Hitler's star rose. He joined the SS and worked for the Gestapo in Berlin. Promotions followed; by the time Norway was being invaded, he'd attained the rank of *Sturmbahnfuhrer* (SS equivalent to major).

On 21 April 1940, he was deployed to Oslo, heading up the newly imposed *Einsatzkommando* (a battalion-sized unit, their allotted tasks usually the mass murder of Jews and any others whose existence offended the Reich). He commanded both the *SD* & *SiPO* and moved up the ladder to *SS-Standartenfuhrer* (battalion commander or colonel).

During the occupation, Fehlis used beatings and torture, and it is estimated he ordered the murder of 151 Norwegians. His zeal did not go unnoticed or unrewarded and in June 1944 he moved up again to *SS-Oberfuhrer*. Von Falkenhorst described him as 'quiet, modest and very polite'. The general was nonetheless aware Fehlis had him and his staff closely watched. At the end of the war, he took cover to avoid retribution from Milorg but, finding the game was up, preferred suicide.

PRELUDE

We come from the land of the ice and snow
From the midnight sun where the hot springs flow
The hammer of the gods
Will drive our ships to new lands
To fight the horde, sing and cry
Valhalla, I am coming...

Black night in a frozen land, the Hardanger Plateau.[1] 19 November 1942, a season Churchill would later describe as 'The end of the beginning' as General Montgomery's Eighth Army completed their defeat of Rommel's Afrika Korps in the Western Desert. But here, in Norway, it was still beginning. The late autumn land was bleak, though a fine night, except cloud was threatening. A deeply hostile environment, even for those hardened Norwegian Commandos of the *Grouse* team. They had already withstood much, testing their toughness, resilience, and stamina to the limit. This was their environment, but it was pushing them to the very end of their endurance. Yet tonight, there was hope.

Hope was embodied in the arrival of a team of glider-born Commandos who would then spearhead a raid on the heavy water plant at Vemork, aimed at striking a precise but decisive surgical blow against Germany's efforts to build the bomb. It was a race the Allies dared not lose, as the consequence of failure could be Armageddon. *Grouse* had been in place for several weeks, now their nerves were tautening as the prospect of action finally materialised.

At 17.15 hours, 2nd Lieutenant Knut Haughland, the team's brilliant radio operator, had reported that weather conditions were favourable over the landing zone. Roughly three quarters of an hour later the assault team took off, two Horsa gliders[2] towed by a brace of Halifax bombers.[3] On board the former were 34 Royal Engineers, trained as Commandos, heavily armed, 17 men per glider including two RAF personnel. It was fair and, for the season, mild. So far so good. For late 1942, this was cutting edge stuff, Combined Operations'[4] first attempt to deploy glider-borne infantry. Theirs would be no easy task, but Operation *Freshman* was now firmly underway.

As the planes towing their heavy burdens climbed into the air from Skitten aerodrome,[5] Team *Grouse* got to work, 2nd Lieutenant Jens Anton Poulsson, with Sergeant Klaus Helberg, began setting up their Eureka machine. An unlovely piece of kit that had already caused difficulties, it was nonetheless essential for comms as the planes carried its other half, named Rebecca.[6] On board the tugs, Rebecca would send a pulse down to earthbound Eureka, which then transmitted back, on a different frequency, enabling crew to identify the drop zone. It was all pretty crude by today's standards but the best thing on offer at the time. Identifying just the right moment to release those laden gliders was a very fine art indeed. This was night, over a strange and silent land, no lights to see and a bewildering tapestry of lakes and rock formations, riven by deep, twisting defiles, none of which was ideal for a first attempt at pretty much anything.

Haughland donned his headphones, snakes crawling in his underfed belly, tension rising as he strained to hear the swelling hum from Eureka. Meanwhile, the team laid out their landing lights – three red in a triangular formation, each 100 yards (91 metres) apart with a fourth flashing white light, their lodestars, at the apex. This was positioned to point downwind – it would flash 'L' for London, angled towards the planes' line of approach. It was around 21.40 hours in that blank shroud of near-Arctic night when both Helberg and Poulsson picked up the noise from an aircraft coming in from the southwest and the hum from Eureka seemed to confirm it. Hopes and tensions rose in tandem. This was what they were here for. Proud Norwegians, they were about to hit back at a detested occupier.

By now the plane seemed just over 1¼ miles (2 kilometres) distant. But then it turned away. It seemed they heard more approaches, but cloud was thickening as hope stalled. At 23.00, they gave up and trudged back to their hut, emotionally drawn, the weight of their physical exhaustion seeming to grow. They remained alone. There would be no raid that night. What had gone wrong? They didn't know.

That was perhaps just as well. The truth would turn out to by just about as bad as they could, in their most dire imaginings, have foreseen. Claus Helberg recalled in a post-war interview:

> On 19 November we were told over the radio that the gliders would be coming the same night. Later in the evening, we went down to the landing site and put out our reception lights. But the wind increased, and it seemed that London had second thoughts. We were informed over the radio that the operation was postponed for an hour. At 2300 hours, we heard the sound of planes. It certainly was a marvellous feat that the pilots had found the place in this weather. The lights were lit, and we expected every moment to see the gliders land. A little later, we heard plane engines, but now black cloud banks were drifting along the mountains, so I assumed that the chances were small for our lights to be seen. The next day, we received a sad report from England.

1

SCOURGE OF THE SWASTIKA

> 'Nilssen. I must tell you; the time has come to make a decisive transition from theory to practice. You've done enough experimentation, my dear fellow. This room has more significance for us than you can possibly realize. Oh, well, one day perhaps he will realize. Don't you think, gentlemen? Forget about fertilizers. This factory is required by the Reich to increase its production of heavy water by 400%. We want 10,000 pounds of heavy water by Easter next year.'
>
> 'That's impossible.'
>
> 'It's not at all impossible, Herr Nilssen.'
>
> *Heroes of Telemark*

Professor Leif Tronstad must have thought that had Norway still been at peace during the autumn of 1941, he'd fully arrived, if only the Nazis hadn't got there first. One of five children, he was largely raised by his mother, his father having died three months after he was born. By 1918, when he graduated from school with top marks in mathematics, his talents were clear. He gained practical experience with a couple of electrical companies before becoming a student at what would be Oslo University's Faculty of Engineering. Having switched to technical chemistry, he graduated first in his year. As well as a budding scientist, he was an accomplished sportsman.

In 1924, he moved up to Trondheim, Harald Hardrada's old capital, to study at the Norwegian Institute of Technology. His work, once again, was deemed outstanding and meanwhile he'd completed his initial compulsory military service, promoted to 2nd Lieutenant armourer. His career flourished, advancing through a series of appointments while finding time to get married and father two children. He worked as an assistant to Herbert Freundlich in Berlin where he made a dazzling breakthrough by divining how to measure very thin oxide surface coatings, a solution which had eluded chemists for a very long period. Next, he was appointed Professor of technical inorganic chemistry at the Norwegian Institute of Technology in 1936, one of the youngest full professors in the country.

> Leif Tronstad was once called the worst student ever accepted into the chemical engineering programme at the Norwegian University of Science and Technology (NTNU) in Trondheim. He turned out to be perhaps its most brilliant graduate. By the late 1920s, the young inorganic chemist used ellipsometry to show some metals had a thin, protective oxide layer. Today, it's a standard technique in thin film chemistry and physics – at the time, it solved why some metals, contrary to their electrode potentials, will not corrode. But Tronstad's name will forever be attached to something else: heavy water.[1]

Tronstad worked with a colleague Jomar Brun,[2] another chemical engineer who was working at Norsk Hydro (from 1929). Brun was a leading player in planning and managing the production of heavy water. The plant at Vemork was already making ammonia to be used in nitrogen fertilisers and, as a by-product, large volumes of electrolyzed water were being generated. Tronstad acted as a consultant for Norsk Hydro. He and Brun were aware of the considerable interest being shown in the acquisition of heavy water, at this stage by both France and Germany (see chapter two).

The professor's tranquil and productive life changed abruptly when the Germans invaded. He was, of course, immediately called up. Tronstad was a true patriot, in the best meaning of the word. Oslo fell before he could get there but that didn't prevent him from organizing local militia to continue fighting. Even when the campaign was over and Norway occupied, he became an early volunteer for a fledgling resistance movement. His cell was based around the Technology Institute, 'Skylark B' was their radio handle. Working still with Brun at Vemork (who was also involved in helping resisters escape by sea), Tronstad began sending London information on the Germans growing interest in the plant. This was risky, so much so that the Professor was obliged in September '41 to make a run for it. He was one jump ahead of the Gestapo and made it firstly into Sweden (where he spurned the lure of a 'cushy billet') and found his way to London.

> In 1940, Germany invaded Norway. Tronstad, a dedicated patriot and officer in the Norwegian army reserves, joined the brief defence. As the Germans occupied his country and the armed forces surrendered, Tronstad took part in the *hjemmefronten* (Home Front) resistance, organising students and staff at NTNU. He was soon living dangerously. In September 1941, he narrowly escaped the Gestapo and fled to London, leaving his wife and two children behind. They corresponded throughout the war via various clandestine operations, but they would never meet again.

Even there, he chose the more demanding option, focusing on fomenting resistance work, rejecting proffered scientific roles. Extremely fit and determined, a brilliant organiser, he volunteered for active service but was deemed too valuable to hazard in the field. He remained very troubled over the Nazis interest in Vemork and its heavy water production. Indeed, as events would show, he had much to be concerned about.

> The Norwegian government-in-exile and royal family had established themselves in London. Here, Tronstad found himself in the middle of enthusiastic patriotism, but also in a complex web of mistrust where everyone could have ulterior motives. The Norwegian cabinet did not trust the resistance; the resistance didn't trust the Norwegian army; the Norwegian

> units in the Allied Special Operations Executive (SOE) didn't trust their British officers. Tronstad was not above motives of his own. Nazi Germany was aiming to develop its own atomic bomb and needed D O from Rjukan to slow down neutrons and sustain a nuclear chain reaction. The Allies couldn't let this happen and so planned to bomb the plant Tronstad had established, but he opposed the move, suspecting it would hand over a world monopoly of D O to Imperial Chemical Industries after the war.
>
> As Tronstad said: 'Blood is thicker even than heavy water!' In the meantime, Tronstad's diary (kept in secret throughout the war) records a chemist with an extraordinary life. He talked with ministers, met with the Norwegian king and crown prince, took night trains up to Scotland to sort things out between the Norwegian SOE troops and their officers, and undertook Commando training. In between, he kept abreast of military-related science developments (always keen on getting Norwegians into the game.[3]

No movie director could have called for a more dramatic setting than Norsk Hydro[4] at Vemork, just outside the town of Rjukan in Tinn. At the time of its construction (1907–1911), and with a capacity of 108 MW, this was the world's largest hydroelectric power plant. Brainchild of Sam Eyde, a pioneering engineer who had begun his stellar career in Germany, its prime purpose was to fix nitrogen for manufacturing fertiliser, based on an innovative method devised by Kristian Birkeland.

An adjacent hydrogen plant was constructed in 1934 and a by-product of production, as noted, through water electrolysis was heavy water.[5] Norsk Hydro were guided in their outputs by a later Nobel Prize winner, Odd Hassel. Nobody then quite saw how significant this might become. (The work for which he is best known and for which he earned his Nobel Laureate established the three-dimensionality of molecular geometry. He focused his research on ring-shaped carbon molecules, which he suspected filled three dimensions instead of two, a common belief of the time.)

Norsk Hydro's development was linked closely to the town itself. A longstanding settlement, the new development between 1905 and 1916 was very much the child of industrial process. It was the nearby presence of the great 340-feet (104-metre) Rjukan Falls that provided water to generate electricity. Despite this industrial pedigree, the landscape is pure Tolkien. The town itself squats in a narrow defile and the encroaching cliffs shut out the sun for long periods, even in the summer (senior executives of the company generally enjoyed living on the sunny side while their less privileged colleagues spent more time in the dark).

The spectacular, tumbling cascade of the Falls had long inspired a well subscribed tourist trade and offered a jumping-off point for skiers who wished to traverse the forbidding fastnesses of the Hardanger Plateau. The plant itself – brutalist concrete and steel – might have been the Black Knight's castle lowering over the valley if it wasn't so purely functional. Not a dragon in sight, though the fire its heavy water might unleash could easily consume whole cities. And it would.

I've been to Norway more than a few times, having family there, mainly in the south at Tonsberg on the Oslo Fjord. My interest was, as a lad, firmly focused on Vikings and I took a rather rackety train of fairly ancient lineage north along the side of the water towards the capital and The Viking Ship Museum on Bygdoy Peninsula. This would be in the late sixties and the place would only have been fully open for a decade or so. At first glance, entering the minimalist space of the great hall the 9th-century Gokstad ship struck me full force. Overwhelming. Delayed by the onset of war, construction had begun some time before. Unlikely anyone involved thought the dogs of war would be unleashed so soon after. In that bright new world of exciting innovation as the plant began life, none could have foreseen how events would shift or how Germany might become so interested in Vemork and for what dark and sinister purpose.

Norway had for some centuries been a part of the wider kingdom of Denmark-Norway. The local economy suffered badly during the Napoleonic wars, caused by British naval blockade. The Swedes had used the pretext to invade but got a bloody nose, yet the Treaty of Kiel in 1814, handed Norway over to their king. Norwegians weren't consulted and thus minded to resist. A compromise was reached, and Charles XIII of Sweden was also recognised as sovereign of Norway. By the latter part of the nineteenth century, significant cracks began to emerge in the fabric of this union. Swedish sabres rattled, but in 1905 the Norwegian parliament voted to dissolve the joint monarchy, instead offering the now vacant throne of an independent Norway to Prince Carl of Denmark, acclaimed by the people as their new ruler, Haakon VII.

In the next decade, lifted by this heady freedom, Norway became a modern, progressive constitutional monarchy, female suffrage, higher education, workers' rights boosted increasing industrialisation harnessing fast rushing waters for hydroelectricity. In 1911, Roald Amundsen reached the South Pole, holy grail of polar explorers, beating Britain's Captain Scott and, unlike Scott's team, getting back safely. What finer way of catapulting a small, recently independent state onto the world stage?

As the Central Powers and the Entente squared up and circled, just waiting for a spark to ignite the fire, Norway remained non-aligned, she just wanted to keep out of it and wisely stuck to strict neutrality. That didn't completely work: their merchant marine during 1914–1918 sailed mainly in support of Britain. Some 2,000 of her crews were lost, due to U-Boat action. Nearly half the fleet was sunk.

Despite this sacrifice, there was to be no peace dividend. The inter-war years were characterised by a flux of economic woes, social issues and unrest, strikes, lockouts and a major recession in their vital fishing industry. Consequently, there was political instability, eight different governments came and went between 1918 and 1935, an average tenure of around 18 months. A split in the more left-leaning parties enabled communists to gain a foothold in the 1920s, but their share of the vote faded away through the next decade. As storm clouds massed again after Munich, Norwegians clung to their tested policy of strict neutrality, but both Britain and Germany had other plans. Hitler was casting covetous eyes and dictators have no respect for neutrals. Norway was perceived as vital to Germany's wider interest in a European war.

It was the Norwegians' collective misfortune that the country had this significant strategic value, linked to Hitler's need to secure his import of Swedish iron ore through the port of Narvik. By the end of March 1940, the Phoney War, this bewildering lull, had endured since the fall of Poland in September. Neville Chamberlain felt something had to be done, the allies had to get aggressive. The French premier, the hawkish Paul Reynaud who had replaced dithering Daladier, was being pushed to do the same. The Allies agreed upon laying mines in Norwegian waters and proposed sending joint expeditionary forces to occupy four strategically vital harbours: Narvik, Trondheim, Bergen and Stavanger. Not untypically, neither could agree on details and this hiatus allowed Hitler to seize the initiative. The allies now couldn't expect to start laying mines before 8 April, by which time German advance units were already en route.

In fact, on 9 April Germany struck first and hard. An unsuspecting Norwegian government was still grappling with the notion their waters, without their agreement, were about to be mined. This soon paled into insignificance as swastikas began waving over key towns and cities. Local forces at Narvik, Trondheim, Bergen, Stavanger and Kristiansand were overwhelmed or overawed. Aerodromes at Oslo and Stavanger were seized in well planned coup de main assaults by paratroops – *Fallschirmjäger*, the first such successful aerial envelopments in history. Any hopes the defenders might have had of holding the capital against amphibious attack were dashed when victorious paratroopers advanced through the suburbs. On the same day, German forces crossed the border into Denmark, which swiftly capitulated. Hitler had stolen a very effective march on Chamberlain and Reynaud.

Despite the suddenness of the blow, Norwegian PM Johan Nygaardsvold and his cabinet colleagues rebuffed a German ultimatum, refusing to accept the 'Protection of the Reich', (surely one of history's most bitterly ironic promises). They swore to fight on, even in the certain knowledge such a struggle was wholly one-sided. Hurriedly packing up, ministers, senior officials, the

King and Royal Family escaped from Oslo by road and rail, firstly to Hamar and then on to Elverum in e municipality of Innlandet County, where an ad hoc parliamentary session was hurriedly convened. Executive power was, as an emergency measure, vested in the King as constitutional head of state, and his cabinet (see appendix 1). Both proved remarkably agile as the Nazi net tightened, before being finally evacuated to London on board HMS *Devonshire* on 7 June.

The Allied riposte ended badly. It wasn't until 14 April that British soldiers began landing at Narvik, then on Namsos and Andalsnes to form a pincer movement aimed at Trondheim. Fresh German forces soon threatened to encircle the Allied bridgehead at Namsos and moved from Oslo against those at Andalsnes. Narvik didn't fall and Germany built up her forces to over 25,000 personnel. In early May, both Allied bridgeheads were evacuated. Britain and France now had even more pressing concerns. The Phoney War was well and truly over, and the Battle for France was brewing. On 10 May, panzers came crashing through the Ardennes Forest, much to the consternation of the French who had considered the difficult terrain impassable by armour. By the end of that month, the BEF was on the beaches at Dunkirk. Norway, for the moment, had ceased to matter.

Even Narvik, once taken, proved a rather hollow victory, any gains soon abandoned in a final British evacuation when the remaining 25,000 soldiers were taken off. Hitler had won the battle for Norway. At a stroke, he had secured valuable strategic bases in Norway and guaranteed Germany's vital supply of iron ore from Sweden. The swastika was being emblazoned across Europe and citizens of occupied lands would soon find out just what that implied.

Enter Vikdun Quisling, the local fascist candidate, leader of the 'National Union' (NS) far right faction.[6] As early as December 1939, Quisling had put himself forward to Hitler as a suitable head of state, should Germany take over. Adolf wasn't listening, but undeterred, on 9 April as the parachutes blossomed over Oslo, Quisling – with no Nazi backing – stormed into NRK studios, delivering an unsolicited broadcast to the nation at 19.30 hours proclaiming himself to be the new PM and ordering all Norwegians to lay down their arms. Nobody did. Even the Germans weren't pleased. They'd have preferred to work with the existing administration, maintaining order and easing the burden of occupation. This changed very rapidly when it became obvious that government wasn't just going to roll over. Hitler, largely because he had no better tool to hand, finally recognised Quisling as new head of state.

Kurt Brauer, the German ambassador, who on the eve on invasion had hosted a jolly where Norwegian guests were treated to a Nazi propaganda film showing the rape of Poland, just to underscore the message that King Haakon, still in the country if not his capital at this point, must recognise Quisling and legitimise his appointment. His Majesty demurred and was rewarded by a visit from the Luftwaffe. The Nazis didn't do subtle or protracted. Watching the otherwise innocuous village where he had sought temporary sanctuary burn, the King's resolve merely hardened.

Suddenly, Vikdun Quisling didn't look such a solid bet after all, and he was brusquely shunted aside. As an interim measure, on 15 April the Germans, established an administrative council under the chair of Ingolf Elster Christensen[7] to manage those parts they'd already seized. Just over a month later, this short-lived expedient was disbanded when *Reichkommissar* Josef Terboven descended as *gauleiter*. Initially, he did attempt to bring the remaining members of Parliament into the fold. He failed. In February 1942, Quisling was dusted off and set up as head of state, though the teeth of the occupation remained Terboven's wolf pack, violence his typical response to any act of resistance.

Control of all the occupying forces, land sea and air remained with *OKW* but everything else devolved onto Terboven as *Reichkommissar.* Legislation legitimising Nazi aims and actions was pushed through, all domestic parties bar Quisling's NS were outlawed, his partisan stooges were foisted on to trades unions and other organisations, not always in the face of opposition. There were many happy to collaborate and others prepared to go along. Membership of the NS did in fact, if not for long, grow during the early months of the occupation.[8]

Having breathed their defiance while still on home soil[9] and safe now in London, King Haakon and his ministers set up a Government in Exile, liaising with the Allies they were far from toothless. They still controlled the country's diplomatic service and vitally, the vast Norwegian merchant marine. They coordinated the nascent resistance movement and formally set aside their cherished policy of neutrality. Norway might be occupied but she was now very much at war. Her merchant navy, with 4.8 million tonnes of shipping, was the world's fourth largest and while the Germans secured perhaps a fifth of the fleet, the rest, as many as a thousand vessels all told, got clear and joined the fight. Their economic contribution, made during the war at high cost, was more than enough to secure the Government in Exile. The very fact the King and executive stayed in being, albeit in London, significantly undermined the dubious legitimacy of Quisling's thug regime.

'Q. Do you know the difference between the Nazis and a bucket of manure? A. The bucket!' Humour is a weapon. After the shock of invasion and occupation, it was the only one the Norwegians had to hand. Over 50,000 had joined Quisling's party, but the inevitable heavy tread of German jackboots rapidly alienated whatever stoicism might have infected the rest.

Most people did just keep their heads down and hope for the best – in 1940 Hitler's juggernaut seemed totally unstoppable. After his legions first sped east following the eruption of *Barbarossa* in June 1941, world domination appeared within his grasp. At the end of that year, before the ice-bound gates of Moscow, momentum stalled. War on two fronts was the very definition of a high-risk strategy. Still, in 1942, the Allies position appeared critical. America might now be in the war, but Tobruk fell in the Western Desert and Singapore in the Far East. Rommel was on the road to Cairo, Hirohito stood before the final frontier of the Raj. Combined Operations' major raid on Dieppe that August, *Jubilee*, was a total disaster.

But there was still humour and with that, hope; 'It's rumoured we're getting new stamps bearing Quisling's likeness, but distribution has been delayed because no one knows which side to spit on!' The crust of resistance, fuelled by SOE, was beginning to harden. Norway might have been an apparently easy conquest but it's impossible ground to police in the face of a hostile populace. By the time Axis forces finally laid down their arms, the garrison stood at a total of 360,000 soldiers. Had those men been available for the Battle of Normandy, the final conclusion might have been very different.

Quisling's parish council might strut their stuff, but Norwegians could see all too clearly where real power lay, vested in the intimidating persona of *Reichkommissar* Josef Terboven. Appointed to his office on 24 April 1940, before the guns even fell completely silent, the *gauleiter* – a career Nazi, moved himself into the Norwegian Crown Prince's official residence and set up shop in the Parliament building, just in case people hadn't got the message. He was answerable to the Fuhrer alone and ruled as a virtual satrap. Small minded, stiff and hubristic, he did nothing to endear himself to his new subjects. But then, his perceived mission had nothing to do with hearts and minds.

Whilst he exercised no control over the bulk of German army personnel, he did directly command a mixed bag of security forces, numbering perhaps 6.000 and 800 or so of these were secret police. Even Goebbels whinged in his diary about Terboven's heavy-handed tactics, the *Reichkommissar* was as unpopular with his fellow Germans. Nonetheless, he found an ally in Wilhelm Rediess the local Higher SS & police commander. Rediess had ample form. At the outset he'd overseen deportation of all Jews from East Prussia, helping keep the Reich racially pure. Around 1,600 of the deportees were deemed mentally ill and therefore in need of purging. He used ad hoc methods such as mobile carbon monoxide units and promising his operatives a 10-mark bonus per murder. It took 19 days. He did not pay the reward.

For all his bombast, Terboven would always play second fiddle to the military governor and conqueror of Norway, General Nikolaus von Falkenhorst. A career soldier aged 55 and from good East Prussian Junker stock, he had joined the Kaiser's army in 1903 and served in a variety of roles throughout the Great War, followed by stints in the paramilitary *Freikorps*,[10] then the more respectable *Reichswehr.*[11] During the 1939 invasion of Poland, he commanded XXI Corps,and on 20 February 1940 that the Fuhrer appointed him to lead a potential invasion of Norway. There was no plan. He was given until 17.00 that same day to construct a proposal. With the aid of a second-hand Baedeker bought that morning, he pulled something together (Operation *Weserubung*) and Hitler gave his stamp of approval. Just like that.

Despite such an improvised beginning he handled the invasion with considerable competence and elan. He suffered losses to his ships during the naval battle at Narvik,[12] but this limited and rather costly Allied victory could not affect the final outcome. Success gained him further promotion (to the rank of *Generaloberst*) and a pretty secure billet for the rest of the war, or at least until December 1944 when he was relieved and relegated to the strategic reserve – but by then time was running out for Hitler's 1,000-year Reich. In late 1942, Falkenhorst had submitted a further plan, Operation *Arctic Fox*, a projected invasion of Sweden but, happily for the Swedes, this didn't proceed; disaster at Stalingrad was already looming.

Any defiance, however outwardly feeble, you throw at an occupying force serves a purpose because occupation doesn't work unless the oppressor attains a level of dominance which eradicates any notion of striking back. You keep that flame alive and you're still in the game. After the shock of invasion and the initial bruising effect of occupation had dissipated and revulsion against Terboven and his swaggering bullies started to harden, resistance began to germinate. As summer turned to autumn in 1940, students from Oslo University could be seen sporting paperclips in their lapels. The humble paperclip became identified as a symbol of unity – 'bound together'. Inevitably, as the Germans recognised these tiny symbols for what they were, they banned such talismans of nationhood, and went on to proscribe others such as the H7 monogram (the King's Royal Cipher) and even red bobble hats!

Underground newspapers began to appear. These illicit presses, with severe penalties for those involved, would not directly affect the occupiers but they acted as sources of information from the free world, countering the Germans' relentless propaganda. Moreover, their very being spurred the Nazis to devote time and resources into tracking them down. Soon, as the tide began to turn, those resources would become increasingly precious. And there was the 'Ice Front', a calculated campaign of baulking the Germans by non-confrontational means, effectively impossible to counter. Don't ever speak to them in their native tongue (though many Norwegians understood German perfectly well). Refuse to sit next to them on public transport – this ostracization from a nation the Germans, at one level, thought of as Aryan, stung to the extent that they banned standing on a bus if a seat was free!

None too subtle pressure was brought on schools to sing to the occupiers' tune. Teachers demurred – and accepted the inevitable harsh consequences. In March 1942, over a thousand who had refused to kowtow were arrested and half of these ended up north of the Arctic Circle as forced labour. Those deemed less culpable suffered more minor sanctions, such as non-payment of salaries. The Nazis were always more at home with the stick than the carrot. (Readers wishing to know more about the invasion of Norway from a Norwegian perspective should refer to a first-rate contemporary account written by the former foreign minister Halvdan Koht, published in England in 1941 – *Norway: Neutral and Invaded*.)

More active resistance began to swell, spurred by the actions of brave men and women on both sides of the North Sea. One of the most inspirational if rather ad hoc responses was the stuff of legend – the Shetland Bus. In March 1943, Leif Larsen, master of the *Bergholm*, a trawler from Lerwick, was undertaking a most dangerous mission against Hitler's *Festung Europa*. His vessel was bound for Traena in the Nordland delivering arms and equipment to resistance fighters based there.

In all, Larsen and his crew were to make over fifty trips, missions fraught with risk from the fury of the elements and the vigilance of the Luftwaffe. This time luck ran out. They made their delivery of vital supplies safely, but motoring back toward Shetland, two cruising ME110s dived and strafed the boat. Of the eight crew six were hit and one, a young lad named Nils Vika, died of his wounds. With the ship sinking, the survivors, mostly injured, took to the lifeboat, and for four days rowed through harsh and unforgiving Atlantic waters. Mercifully, they were rescued off Alesund by an MTB from Shetland.

Larsen was already something of a legend. He had escaped from occupied Norway in February 1941 in the fishing smack *Motig 1* and, after initial training, became a skipper on what was to become known as the 'Shetland Bus'. This was largely unsung but highly dangerous work, scratching at the flanks of Fortress Europe. Such brave men in small boats undertook a constant series of daring missions, sea passages contesting the worst of the elements and defying the fury of the Nazis. The German invasion and subsequent occupation of Norway in 1940 is chiefly remembered for the fiasco of attempted Allied intervention. But thousands of Norwegians, finding the yoke of Nazi tyranny intolerable, escaped in fishing boats and other small craft, many landing in Shetland.

Those who remained, in many instances, formed the core of a nascent resistance movement. By summer the following year, a base had been established at Lunna on the north-east flank of Shetland mainland and civilian craft were pressed into service as transports for supplies and agents being infiltrated into Norway. Thus, the Shetlanders were at the forefront of a small but steady assault on Hitler's Europe. Many of those who fled this grim oppression did so in conditions of extreme peril. Kare Iversen, a fisherman who joined the Norwegian resistance in 1940, pressed his father's craft *Villa 1*, a 13-metre fishing vessel, into service, when he and several companions, hunted by the Germans, sought refuge in Shetland. Their passage was eventful:

> The first two days passed in perfect weather conditions. But at 3.10 on our third day trouble arrived. I was coming out

> of the engine room when I saw through the wheelhouse door that a German flying boat was coming straight for us. At the time two of us were down in the engine room, two in the forward cabin. After the first burst of gunfire, I went out on deck and released all the halyards to let the sails drop to the deck, then we went back down to the engine room. The Germans continued shooting at us for twenty minutes. Their gunfire riddled the wheelhouse and holed the boat just above the water line with their shelling. Whenever we rolled, a big rush of water now came into the hold. The flying boat tried to land but the sea was too rough, so they gave up the attempt.[13]

After a fraught passage in high seas, his vessel damaged by Axis gunfire, Kare reached Shetland, anchoring at Sandwick. After initial de-briefing, he trained with fellow Norwegian exiles and then returned to Lunna where he sailed with Leif Larsen on *Arthur, Feioy, Siglaos* and *Heland.* In December 1943, after many adventures, he became part of the crew of MTB *Hessa* and served until the end of the war, though he found time to court and marry a local girl, Cissie Slater.

Lunna was a reasonable beginning, but the place lacked adequate facilities, so by 1942, efforts had been switched to Scalloway where a local concern, William Moore and Sons, acted as engineers and repairers for the growing fleet of small boats. This was the very stuff of derring-do, a channel for the frustrated exiles, with local assistance to strike back at the occupier. It was not undertaken without cost; *Sjo, Aksel, Feioy, Bergholm* and *Brattholm* were all lost with the lives of over thirty men. As unit strength was never more than double that number, such losses were dismal indeed.

When it was suggested that this was unacceptable and that the sailors could be more effectively employed in a conventional naval role, they simply replied, 'Give us better boats and we will continue the fight.' This proved a turning point. The Shetland Bus service was re-equipped with three US fast motor launches, *Hitra, Hessa* and *Vigra.* These sleek craft were built entirely for war, 100 feet in length and powered by twin 1,200 hp General Motors engines. Cruising at 17 knots with a further five in reserve at full throttle and bristling with machine guns, they could give a good account of themselves against marauding Messerschmitts.

From November 1943, they replaced the ad hoc conversions to carry out many further operations. None was lost, nor did the crews sustain any fatal casualties.[14] Shetland had been in the front line since November 1939 when half a dozen Heinkels came in low over Lerwick, aiming to attack vessels in the harbour. Eight bombs were dropped to no effect, but strafing attacks wrecked a flying boat, though all the crew were rescued by a brace of haddock boats. Happily, three trawlers each heavily loaded with marine munitions, in this case depth charges – *Northern Foam, Northern Isles* and *Northern Princess* – were spared. Once Norway fell, there was a possibility that Shetland could form the target beachhead for an invasion of Britain.

When Germany had invaded in 1940, the rationale was firstly to secure her vital stocks of iron ore coming out of Sweden, her prime supplier, going overland to Narvik and then on by sea. She could not allow that precious umbilical to be severed by Allied action. Secondly, the long, serrated coastline offered a handy refuge for seaborne predators and would frustrate any attempt by Britain to blockade Germany, as she had done so effectively in the previous war. At that stage, the Vemork plant with its heavy water production, whilst certainly on the Axis radar, wasn't the prime reason.

This would change.

2

THE URANIUM CLUB

'We're taking over this ship in the name of the Royal Norwegian Government. This ship's going to England.'

'Like hell it is!'

'You've got a choice. Sail under the King as free men or be prisoners for the rest of the war.'

Heroes of Telemark

don't know what's happened:
the TV is blank, WhatsApp goes, alarms go off, some glass shatters
but there's no great explosion. Not here. Not where you are.
Just acrid smelling rain – and cars that screech
into a neighbour's drive – man screaming, 'a bomb went off!
It's all gone! The whole lot has gone!'

J. A. Hartley, 'After Party' (2022)

In the summer of 1861, the Union fell apart and the US descended into civil strife. Some confederate volunteers marched to war hefting flintlock muskets their grandfathers had carried in the war of 1812 and very often used by *their* fathers in the Revolutionary War. By the time the Civil war ended four years later over the prostrate corpse of the south, Union troops were deploying magazine-fed breech-loading, lever action repeating rifles – modern firearms in short. In those four bloody years weapons technology leapt ahead more rapidly than it had done in the preceding century. Industrial warfare had arrived; the body count soared. That was just a beginning, it would get much, much worse.

One of the great innovators was William Lord Armstrong (1810–1900) who moved from provincial law practice into the science of hydraulics, then onto arms development and manufacture, a pioneer in the evolution of rifled, breech-loading artillery. In his youth a 6-pounder might weigh 1½ tons – by the end of his long life, his Elswick Yard was turning out massive naval guns weighing a hundred times more.

Tommy Atkins' grandfather who, in his redcoat, marched along the hot pavé would have easily recognised his own grandfather toiling over the same roads under Marlborough, same uniform, same weapons and only moving as fast and far as he could march in a day. Neither would have understood Tommy's war in 1914 where trains could shift huge quantities of men, horses and materiel, where quick- firing artillery would dominate battlefields, dug in and covering 25,000 miles of trenches in a four-hundred-mile front, where instead of two rounds a minute, barely accurate at 50 yards, the machine gun could spew out 500 bullets and kill you at 2,000 yards.

Perhaps 10 million people died in the Great War, the next one consumed at least five times as many. Nineteenth- and early twentieth-century science fiction writers like H.G. Wells had prophesied an age of new and more powerful weapons,[1] armoured leviathans rumbling over No-Man's-Land, crushing all in their wake. In 1916, that is precisely what happened.[2] Armageddon was also a concept that seemed to reside only in apocalyptic fiction. The bombings of Hiroshima and Nagasaki on 6 and 9 August 1945, would claim up to 200,000 lives. Most in an instant. Armageddon suddenly came of age and the world had shifted. Norwegians in 1940 had no idea that Vemork would come to play such a perceived pivotal role in a race for the bomb.

The first successful device, code name Project Y, detonated by plutonium was exploded at a test site south of Albuquerque in New Mexico on 16 July, less than a month before US B-29 'Enola Gay' dropped an A-Bomb on Hiroshima. This test blast was the successful outcome of what had been codenamed the Manhattan Project, based at a top-secret site at Los Alamos National Laboratory, also in New Mexico. The project was supervised and managed by Brigadier Leslie R. Groves with the brilliant physicist Robert J. Oppenheimer. When the bomb fell on Hiroshima it exploded with the force of 15,000 tons of TNT and obliterated 4½ square miles (11.4 square kilometres). If anything could be worse than the blast, it was the after-effects of radiation, which cursed not just survivors but their children. A weapon of biblical terror and hideous lingering effect.

The key to the Bomb is nuclear fission, uncovered on 19 December 1938 by the German chemist Otto Hahn, working with Fritz Strassman and Austrian-Swedish physicist Lise Meitner. Hahn has since been referred to as the father of nuclear chemistry and nuclear fission who also revealed the phenomena of atomic recoil, nuclear isomerism and rubidium-strontium dating. Born in 1879, a graduate of Marburg University, he subsequently studied under Sir William Ramsay[3] and then with Ernest Rutherford[4] at Montreal's McGill University. Hahn returned to Berlin in 1906 where he was offered an old woodworking studio in the basement of the university. From such a modest start he was within six years heading up the Radioactivity department within the prestigious Kaiser Wilhelm Chemical Institute where he began his groundbreaking collaboration with Lise Meitner. He joined up as a humble *landser* in 1914 but served under Fritz Haber[5] engaging in the new devilry of chemical warfare. He served across all of the major fronts and came out with an Iron Cross 2nd Class.

After the war he returned to academia and the chair of the Chemical Institute. During the early years of Hitler's Third Reich, he worked closely with Meitner and Strassman, studying isotopes created through neutron bombardment of uranium and thorium, which culminated in the great breakthrough. Never a Nazi, he opposed the virulent Anti-Semitism which consumed Germany like a cancer after 1933. This was, in terms of academe, hugely self-defeating as many Jewish scientists, like Meitner, were obliged to flee and seek sanctuary abroad. Nonetheless, during the war Hahn worked on the development of Germany's nuclear weapons.

Lise Meitner, a year older than Hahn, discovered radioactive isotopes of radium, thorium, protactinium and uranium, Einstein dubbed her 'The German Marie Curie', an unintended irony given that her faith excluded her from the master race and, had she not fled, might have expected to end her days in a death camp. She successfully completed her doctorate at Vienna in 1905 but spent most of her academic career at the Kaiser Wilhelm Institute – the first woman in Germany to hold a chair in physics – until the Nuremburg Laws. She left for Sweden and Hitler lost one of his finest scientific intellects.

Franz Strassman, a generation younger than Hahn or Meitner, came from a humbler background, growing up as one of nine children in Dusseldorf. His father died when he was still young and poverty restricted his career choices. Still, he enrolled as a chemistry student at Hannover's Technical University in 1920, finally gaining his doctorate nine years on, confirming his abilities as an analytical chemist. Next, he was granted a partial scholarship at the Berlin Institute where he first encountered Otto Hahn, who took him on as an assistant. Strassman was also an accomplished violinist. He was no fan of Adolf Hitler. He resigned from the Society of German Chemists once this was subsumed into the sucking maw of the fascist state. In retaliation he was blacklisted – effectively excluded from the state sector.

Meitner came to the rescue and persuaded Hahn to create a position which rescued him from oblivion. The Nazis didn't win his heart or mind: 'Despite my affinity for chemistry', he wrote, 'I value my personal freedom so highly that to preserve it, I would break stones for a living.' During the war, he and his wife Maria provided a refuge for Jewish musician Andrea Wolfenstein, despite the risks to their family. He continued to work in the field of radiochemistry but would have no part in weapons technology, vowing he'd rather commit suicide than help Hitler build the Bomb. It is one of history's ironies that these three brilliant minds would unlock the key to the process that would facilitate the creation of nuclear weapons and give Hitler his start in the race.

In the course of researching this period, I have read a lot about nuclear fission and failed to understand most of it. But, in a nutshell, nuclear fission is a reaction in which the nucleus of the atom splits into two or more lesser nuclei. This process of fission can create gamma photons, releasing great amounts of energy. This energy fuels nuclear power and drives the blast of the Atom Bomb. Nuclear fuels undergo the fission process when they collide with fission neutrons and emit further neutrons when they fracture and split. This makes possible a self-sustaining nuclear reaction and this reaction releases energy in a controlled sequence within a reactor or at a vastly more rapid and uncontrolled rate in a nuclear weapon. Hahn and Strassman in their second paper introduced the existence and liberation of additional neutrons as part of the overall fission process, leading or pointing to the nuclear chain reaction. It was this which led on 6 August six years later to the obliteration of Hiroshima in a heartbeat.

If Hahn, Meitner and Strassman weren't too keen on Nazi ideology, one of that younger generation who had no qualms whatsoever was Kurt Diebner. Born in 1905, he studied physics at Martin Luther University in Halle-Wittenberg, graduating in 1928, then completing his higher degree two years later at Leopold Franzens University of Innsbruck. He returned to Halle to complete his PhD under Gerhard Hoffman, a distinguished nuclear physicist and future member of the Uranium Club. Diebner fitted effortlessly into the twisted hierarchy of the Third Reich, advisor on nuclear physics to the Ministry of Defence (latterly Reich Ministry of War) and the Army Ordnance Office.

Before the war, on 22 April 1939 Wilhelm Hanle, leading German experimental physicist, delivered a paper on the use of uranium fission in a nuclear reactor. Hanle with Georg Joos, also an experimental physicist and influential author, alerted Wilhelm Groth (whose principal field was the development of centrifuges for enriching uranium), at the Reich Ministry of Education ('REM'), to potential military applications. Within a week Wilhelm Dames from REM chaired an ad hoc committee which numbered Hanle and Joos among its members, to consider possibilities. As a result, an informal working group came together at Georg-August University in Gottingen. This pioneering team formed the first Uranium Club (*Uranverein*). In due course, excitement mounted as the Army Ordnance won the turf war and took over from the Reich Research Council of the REM to establish a more formalised German Nuclear Energy Project – the second *Uranverein*.

Formally instituted, fatefully perhaps on the day German forces rolled into Poland, this invigorated body first met on 16 September 1939, chaired by the ambitious Diebner, who was then director of the Nuclear Research Council headed up by General Carl Heinrich Becker. From the start, total subservience to Nazi ideology, narrow, vicious and dogmatic, meant that many of the finest scientific minds – Einstein and Meitner being just two – had fled the authoritarian diktat of the Reich. Others like Hahn were disdainful of what they perceived as less talented, definitely less scrupled parvenus like Diebner. Those Jewish academics and scientists who hadn't already voted with their feet, were rigorously purged. Pragmatism couldn't be allowed to stand in the path of Aryanisation. Many of those who might have contributed were simply called up, now squaddies instead of scientists.

Over time, the project pursued three main strands: (1) developing a nuclear reactor, (2) uranium and heavy water production with (3) uranium isotope separation. The team involved had impressive credentials: Diebner himself with Abraham Esau, Walter Gerlach, and Erich Schumann, serious talent. Even Hahn, despite his disdain for Diebner, was drawn into the mesh. Despite this impressive line up, the project, in terms of nuclear weapons, never quite delivered and enthusiasm began to wane. The Allies could not know this, but they were way ahead of the game.

As early as January 1942, Army Ordnance was beginning to lose interest and passed the buck to the Reich Research Council (while continuing to foot the bill). Split up into nine separate departments, each chasing its own director's objectives, impetus was lost. In retrospect, the scale of effort and resources put in by Britain, the USA and Canada, dwarves the Nazis' provision. The Manhattan Project was instigated by a small but very focused group of scientists who shared a common goal, government in time bought into the idea and offered total support, backed by almost infinite resources. The Bomb ceased to be a theory. Even then, the war in Europe was over before the device could be deployed. In 1940, this reassuring hindsight wasn't available.

Leading German historian Klause Hentschel observes that the Allies Manhattan Project remains a prime example of what he terms 'big science'; a united, concerted and thoroughly resourced team, and entirely dedicated. He contrasts this with the divided and divisive Uranium Club where personal and departmental jealousies clouded the objective. The Manhattan Project's costs were a thousand times greater and employed some 120,000 people at its peak; add in the supply chain and the work involved nearly one per cent of America's vast labour force. There was really never any comparison.

Leif Tronstad had, quite innocently and inadvertently, contributed to the Allies' problem. Nature had created the essential scenario at Vemork with millions of gallons of water streaming down from the hard plateau and collecting in the handy bowl of Lake Mos. For a distance of 18 miles (30 kilometres), the fast-flowing River Vana swept the water down through the slash of Vestfjord Valley, finally debouching into Lake Tinnsjo. Norsk Hydro re-engineered God's bounty into 145,000 kilowatts of electricity by damming Lake Mos and sending the outflow through three miles (five kilometres) of underground channel to the power station at Vemork. Eleven steel pipelines directed the flow down through a drop of 920 feet (280 metres) into the turbine generators. It was a truly Olympian feat of engineering that would surely have astonished the old Norse gods who'd watched over the Hardanger for so many millennia. When this all began in 1906, nobody was thinking of any application to weapons of mass destruction.

Of this huge flow, some 16 tons (16,000 kilograms) an hour was diverted into a hydrogen plant which clung to the edge of the precipice. Here, it was directed into many thousands of steel electrolysis cells which nonetheless consumed virtually all of the Herculean power created by the station. Electricity which ran through these cells split the two atoms of hydrogen from the single oxygen. These resultant gases were in turn pumped down into a series of chemical plants in the town of Rjukan below. All of this colossal investment and technology was intended solely to produce hydrogen for commercial fertiliser. The market demand was sufficient.

In the basement section at Vemork a small portion of the water was further refined, employing a cascade of specialised electrolysis cells which reduced and reduced down to a diminutive but steady dribble. This further refined product was heavy water. In the 1930s this was still cutting edge, the product hadn't been recognised until US chemist Harold Urey won a Nobel Prize for its discovery at the beginning of the decade. It was Tronstad, working with his buddy Jomar Brun, who had the idea of turning this by-product into a commercial opportunity. Tronstad didn't really have much of an idea as to what that opportunity might look like, but he was a great exponent of developing the product then seeking a market.

These two gifted friends and collaborators had designed a process to streamline heavy water production and it is ironic that it was their efforts which would later so alarm them both and lead Tronstad to the conclusion this production facility had to be put out of action. At the same time the Allies could not have sought or found a more able and experienced agent. Who better to wreck something than the fellow who created it?

Tronstad and Brun's apparatus has been likened to a number of cans, stacked into an inverted pyramid. The uppermost row is comprised of 1,824 electrolysis cells which refine the water, (potash is added to promote conductivity). The whole system is designed to create a process of continuous refinement in motion, distilled through decreasing rows of cells. The real magic doesn't occur until the product, still only ten per cent imbued with deuterium-rich water, descends into the bottom 'can'. This was the 'high concentration' phase. Seven distinctive electrolysis cells are set up in line abreast. This is also a further cascade which recycles the deuterium as a gas back into the overall process. Heavy water concentration now rises between the cells with the final dripping product purified to 99.5 per cent heavy water.

This was technological innovation at its best and recognised as such internationally. But so what? Who wants to buy the stuff? It clearly possessed applications or the potential for them. Tronstad was convinced it would have an important role to play in the new dynamic field of atomic physics and for medical research. That the potential did indeed exist was generally recognised but that still didn't morph into a full order book. Labs across the world ordered product but in very small amounts. By 1938, Vemork was selling 80 kg a year, in the money of the day no more than $40,000 in turnover, about £8,000 in sterling (say £450,000 today). In 1939, on the eve of war, the 'suits' at Norsk Hydro had concluded that the game wasn't worth the candle and production was mothballed.

That would change. In Germany, while Diebner with his Uranium Club marks one and two, might never build a bomb, they'd found a use for heavy water. Werner Heisenberg, certainly no admirer of Diebner, was generally seen as the leading exponent of theoretical physics, (now that Einstein had been driven into exile). His assigned task was to work on chain reaction theory. He made good progress towards building a machine that would employ uranium and a moderator to generate reliable levels of power. The question was what to use as a moderator – this is the material set within the reactor which slows down high-velocity neutrons to increase the probability of achieving fission. By the time Falkenhorst was jotting down his notes for the invasion of Norway, Diebner's team had concluded that heavy water offered the best option. A market had just emerged, if not the one Tronstad had been hoping for.

Commander Eric Welsh was one of that eclectic caste of weird and wonderful eccentrics the war threw up, brilliant, erratic and contrary by turns. He was a naval officer, intelligence chief and chemist. During the previous conflict he had worked in marine paints and since 1919 had been with International Paint Limited in Bergen. He spoke the language fluently and was a natural choice to head up the SIS Norwegian desk from 1941. He was considered something of a ladies' man, a heavy drinker and relentless practical joker. Idiosyncratic, he didn't tolerate fools and was possibly one of the role models for James Bond. He made friends and enemies in equal measure.

A recent biographer noted, 'There can be few intelligence officers who have divided opinion among colleagues but who have achieved so much as Erich Welsh. Variously a member of the RN, RAF and SIS and a chemist working on maritime paint, Welsh seems to have employed a personal approach to all aspects of his work.'[6] His libertine lifestyle led to deteriorating health and contributed to his early death in 1954.

Still, as Skylark 'A' to Tronstad's Skylark 'B', Tronstad spirited the mettlesome professor out of occupied Norway and the two met at the suitably peculiar St Ermin's Hotel, Westminster in the same week as Tronstad arrived.[7] They were an unlikely pair: Tronstad the lean, smart sportsman and academic, Welsh, short, portly, unkempt and frequently decorated with cigarette ash. He was no dilettante. He had bravery medals from the previous war and now he owned an asset codenamed 'Griffin' – in fact, a scientist named Rosbaud[8] who operated in proximity with Hahn and other leading German physicists. Rosbaud picked up vibes about Germany's nascent atomic weapons programme even though, once under the Army ordnance banner, doors were closed and mouths. Still, this put him on the same wavelength as Tronstad.

Britain's quickening interest in atomic hardware was the province of the Directorate of Tube Alloys, formerly MAUD,[9] and sounding like something from a Bond novel. It was the fiefdom of Sir Wallace Alan Akers,[10] a graduate of Christ Church where he'd read physical chemistry. His first class honours degree enabled him to get a job in industry, spending some years in the Far East, returning in 1928 to join ICI and rising to manage their Billingham operation on Teesside before being reassigned to ICI HQ. By now, the need to re-arm was obvious and pressing. By 1940 it was blindingly obvious that too little had been done and too late. In the course of 1941, he was recruited by government to head up Tube Alloys.

This was a flagstone in the path leading to the Manhattan Project. Akers' commanding mien and energy were defining characteristics, but his corporate background didn't necessary endear him to the US allies, who divined he had an eye to post-war commercial possibilities. James Chadwick (who had won a Nobel Prize in

1935 for discovering the neutron), was felt to be a more acceptable choice to head the UK mission to the Manhattan Project. Aker hung on to his turf at Tube Alloys. Welsh was the linchpin that would hook Tronstad into the inner workings of UK atomic weapons policy. Vemork would soon be on numerous agendas.

Hitler had boasted of his weapons of mass destruction even though, in 1939, he had none. But Adolf's posturing couldn't be ignored. After all, by the summer of 1940 he had conquered most of Northern Europe, a whirlwind few previous conquerors could match. Anybody that unstoppable had to be taken seriously. At the time of writing, Russian President Vladimir Putin is making similar threats and nobody dares disregard these either. One who took the Fuhrer very seriously was Sir Henry Tizard, head of the Air Ministry's Research Department.

Sir Henry was something of a phenomenon himself, chemist and inventor who had started WWI in the Royal Garrison Artillery (RGA), before transferring to the Royal Flying Corps (RFC). As a scientist, his mantra was all about asking the right question, identifying the problem clearly before embarking on finding a solution. His eyesight wasn't good, but ultimately he took to the skies, acting as his own test pilot in his role as the Corps' experimental equipment officer – aeronautics becoming his specialism. By 1918, while still in uniform, he was working in the Ministry of Munitions.

During the interwar years he pursued a successful academic career, ending up as Rector of Imperial College. He was one of those who began the search for radar and by 1940 was effectively the Air Ministry's chief scientific advisor. On 19 March 1940, his in-tray included a report by two young émigré Jewish physicists Otto Frisch[11] and Rudolf Peierls[12] entitled 'On the Construction of a Super Bomb'.[13] In this paper they argued that just over a pound of pure U-235, split into two or more components which were then flung against each other at tremendous velocity could trigger a blast of such power it could wipe out an entire urban centre, generating cataclysmic temperatures. Clearly this wasn't good news and when they suggesting the Nazis were on track to develop such a device, they had Sir Henry's full attention.

The knight didn't need telling twice and his voice counted. Churchill listened. Within a month, MAUD was set up with Akers in the driving seat and the finest minds British government could assemble went to work. Inspired by Frisch and Peierls, whose credentials were impeccable and whose inside track on German thinking invaluable, the Committee swung into action. Nobody needed reminding of the urgency or potential scale of the threat from Berlin.

Oslo is perhaps an unlikely setting for spy fiction and forties *noir*. The pre-war population was probably not above half a million, the streets wide and clean with stolid civic buildings of modest scale, certainly not Paris, Prague or Berlin. A pleasing place if perhaps rather provincial compared with its diverse European neighbours. The stirring events there just before the clouds darkened could have come from the pages of Buchan or Erskine Childers. The daring mission to lift heavy water stocks under the noses of the omnipresent *Abwehr* was wholly a French project.

France and French scientists have felt rather neglected in terms of their contribution to nuclear physics and rightly so, as Marie Curie and Henry Bequerel were noted pioneers, It was Madame Curie's final assistant Bernard Goldschmidt who came to be thought of as the father of French nuclear weapons. The Curie family would net a world-beating total of five Nobel Prizes. Marie's daughter Irene married another brilliant young physicist, Frederic Joliot, and together the Joliot-Curies clinched the 1935 chemistry prize for their joint discovery of induced radioactivity. On 14 February 1940, Frederic had a date at the prestigious Hotel Majestic on Paris's Rue la Perouse. This wasn't a Valentine's Day assignation; he was to sit down with France's Minister of Armaments Raoul Dautry and the super-suave banker Jacques Allier. No flowers.

Anyone who enjoys Alan Furst's superb range of spy novels set partly in this shadowy phoney war era would instantly see Allier as a classic protagonist. His age ran with the century, and he'd studied at the French Political Science Academy.[14] Armed with a law doctorate he'd joined the *Banque de Paris et des Pays-Bas* in 1923, rising swiftly through the ranks. He was a natural recruit for the *Deuxième Bureau*, France's Secret Service. Immaculately tailored, worldly and expansive, many would have been surprised if he wasn't a spy. The war was in limbo, even if many, in dread of what might be, had already abandoned the French capital, which had lost its brilliant pre-war lustre. Poland had been crushed and surely Hitler would resume the offensive – a repeat of 1914, this time with the Luftwaffe as an overture. The Allies were just holding their breath.

Joliot-Curie explained he was working on building a machine that could unleash energy stored within the atom, in his case aimed at creating revolutionary motors for submarines, but the process did have potential to deliver massive doses of explosive on a scale never seen before. He used the disarming and disturbing analogy of a kitchen table; if the pent-up energy in this bland domestic piece of furniture could be unleashed, there was the potential for an apocalypse. He had Allier's full attention. Dautry, too, was already a convert. The boffin needed heavy water and there was only one supplier, Norsk Hydro. Allier's daytime employer, the *Banque de Paris*, was a major shareholder in the Norwegian company, sufficient leverage surely to obtain whatever product there was? This initiative had already been signed off not just by the minister but by Daladier himself. It came from the top.

Allier needed no further prompting. Furthermore, he had intelligence of his own: on a recent visit to the Bank, Norsk Hydro's chief legal advisor, Bjarne Eriksen, had intimated the company had received approaches from Berlin. The Germans were trying to buy up as much heavy water as they could lay hands on. Norsk Hydro's failed venture was suddenly getting an infusion of new life – a very substantial infusion, these German buyers were suggesting they'd like to acquire as much as two tons of product, an unprecedented amount. At present this level of supply wasn't possible but the Axis buyers were persistent, just as they were very cagey about just why they needed so much.

As for the banker turned spy, the brief was right up his street. He was to trump the Germans by snatching every drop of heavy water the Norwegians had in store and bringing the whole cargo back to France or, failing that, he was to sabotage the lot. Suddenly, this war wasn't phoney at all. Barely a fortnight later, Allier was on the train to Amsterdam, he travelled under false papers made out under Friess, his mother's maiden name. In his pocket a letter of credit from the bank for up to 1.5 million krone and carte blanche to commandeer the services of any French official, or less than official, agent he needed.

In this twilight limbo before blitzkrieg, it was possible and easy to travel through neutral states as if there was no war. Allier flew from Amsterdam to Malmo and from there he travelled by rail to Stockholm. In the Swedish capital he briefed a trio of spies, setting up an RV is Oslo. On 4 March, his train rattled into the Norwegian capital's Central Station, a sober buff-coloured exterior with just enough flourish to make it interesting and went directly to the French legation. Here he learned through radio intercepts that the *Abwehr*[15] were on his tail. They knew all about his fake ID, had guessed what he was likely up to and aimed to stop him. The *Abwehr* needed to be taken seriously. Their means of stopping him could be permanent.

Allier proved hard to scare. He contacted Norsk Hydro from a public phone booth and was immediately admitted to corporate HQ, only a stone's throw from the palace itself. When he met Dr Axel Aubert, the company CEO, Allier had to seem rather abstruse, if obtuse, when saying why he so suddenly needed to purchase Norsk Hydro's entire stock of heavy water. He was pushing on an open door. The Norwegian's sympathies lay firmly with France and not the Nazis. He could have whatever he needed. The very next day both men were on their way by car to Vemork. Their visit was essentially clandestine. Allier could never ignore the fact his enemies were just one step behind him. For both sides, the stakes were very high indeed.

At Vemork, Allier and Axel Aubert met with Brun. Neither at that stage would have known that Brun, of all people, could be trusted, but Aubert, as his boss, wasn't offering explanations. The French were to have the full 185-kilogram supply, all Vemork's current stock with no questions asked. The heavy water would be placed in 26 bespoke stainless steel flasks that could fit into suitcases. Brun was nobody's fool and would have quickly understood what all this was about. The whole cargo would be moved to Oslo by truck ready for onward transmission.

Aubert didn't even want the cash; he was happy to offer the heavy water on loan to France and to undertake a solus supply agreement to cover future production. Allier would owe much to both him and Brun, who swiftly sourced a local sheet metal worker to fabricate the flasks. By 9 March, two truckloads had left Vemork. With Brun riding shotgun the flasks were safely delivered to a safe house in Oslo. Allier had almost effortlessly accomplished the first part of his mission. He had secured the whole supply of Norway's heavy water. All he had to do now was to get the booty safely back to France. He could safely assume the Germans would strain every sinew to relieve him of this precious cargo and, neutrality be damned. It was no holds barred.

Time was of the essence; using a French submarine sliding up the Oslo Fjord was attractive, but with his triumvirate of agents already in place, a swift dash by air was preferable. The Germans would, it was true, be waiting for just this, but his plan was to use deception and a decoy. Two planes were schedule to take off from Fornebu aerodrome on the morning of the 12th. One was to go to Amsterdam and the other to Perth. Allier's agents booked tickets on both, using aliases, spreading their purchases via several travel agents. Just to further muddy the frothing waters, they bought additional seats on further scheduled flights over the two consecutive days.

Allier now staged a theatrical performance worthy of Ian Fleming at his most inventive. Early that chill morning he and one of his stooges, Fernand Mosse, took a taxi to Fornebu, kitted out as flash executive types and making sure everybody in their vicinity was aware of their destination – Amsterdam. Just in case any watchers hadn't got the message, they deposited their luggage for loading in the shape of several well laden suitcases. Both planes were JU52s and standing nearly wing to wing on the apron. Soon the Amsterdam flight was ready to board and propellers were starting to turn. Deus ex machina, a second taxi spurts over the tarmac. Jehan Knall-Demars, another of Allier's team, begged gate staff to allow the car onto the runway so he wouldn't miss his flight. They acquiesced and he ordered his driver to stop the vehicle in the gap between aircraft, a space invisible to anyone watching from the terminal.

In the boot were several suitcases holding thirteen of the precious flasks. These were rapidly thrown into the hold of the Perth shuttle. Allier and Mosse discreetly switched planes as Knall-Demars exited the terminal, crouched low in the taxi he had arrived in. Did the ruse work? Well, two ME109s forced the Amsterdam plane down, casually violating international air space, obliging the pilot to touch down at Hamburg. The *Abwehr* discovered they'd impounded a cache of second-hand suitcases filled to bursting with bricks. On the 13th, Jehan Knall-Demars boarded another scheduled flight to Perth bringing the rest of the cache with him. Round one was a triumph of impromptu ingenuity and swift movement for the Allies. Germany was denied a ready supply of heavy water.

But this contest was only just beginning.[16]

3

THE MINISTRY OF UNGENTLEMANLY WARFARE

> 'The scientific and military consensus in America is that the Germans may be ahead of us in the race to achieve controlled atomic fission. If they are ahead and they get their atomic bomb, they've won the war.'
>
> *Heroes of Telemark*

> 'Taint what we have, but what we give,
> 'Taint what we are, but how we live,
> 'Taint what we do, but how we do it,
> That makes this life worth going through it
> Resolve each day to perform what you ought.
> And to perform without fail what you resolve.
>
> J. White (143 Battery Field Artillery)

I'm a collector of Fairbairn-Sykes (F-S) knives and William ('Dangerous Dan') Fairbairn is a long-time hero of mine. Born in the south, he spent much of his youth in Northumberland in the tiny village of Lesbury. Lean and wiry, he was never the Arnold Schwarzenegger breed of muscle-bound enforcer, just a lot more dangerous and very much the real thing. He lied about his age to join the Royal Marines at sixteen and spent four years, 1903–1907, in the Far East, based in Korea, then occupied by the Japanese (our allies at that point). He became gloriously obsessed by martial arts enlivened with bayonet fighting. He spend his spare time sparring with Japanese opponents in 'friendly' bouts and, to be fair, the Japanese weren't frightened of using bayonets and very enthusiastic exponents.

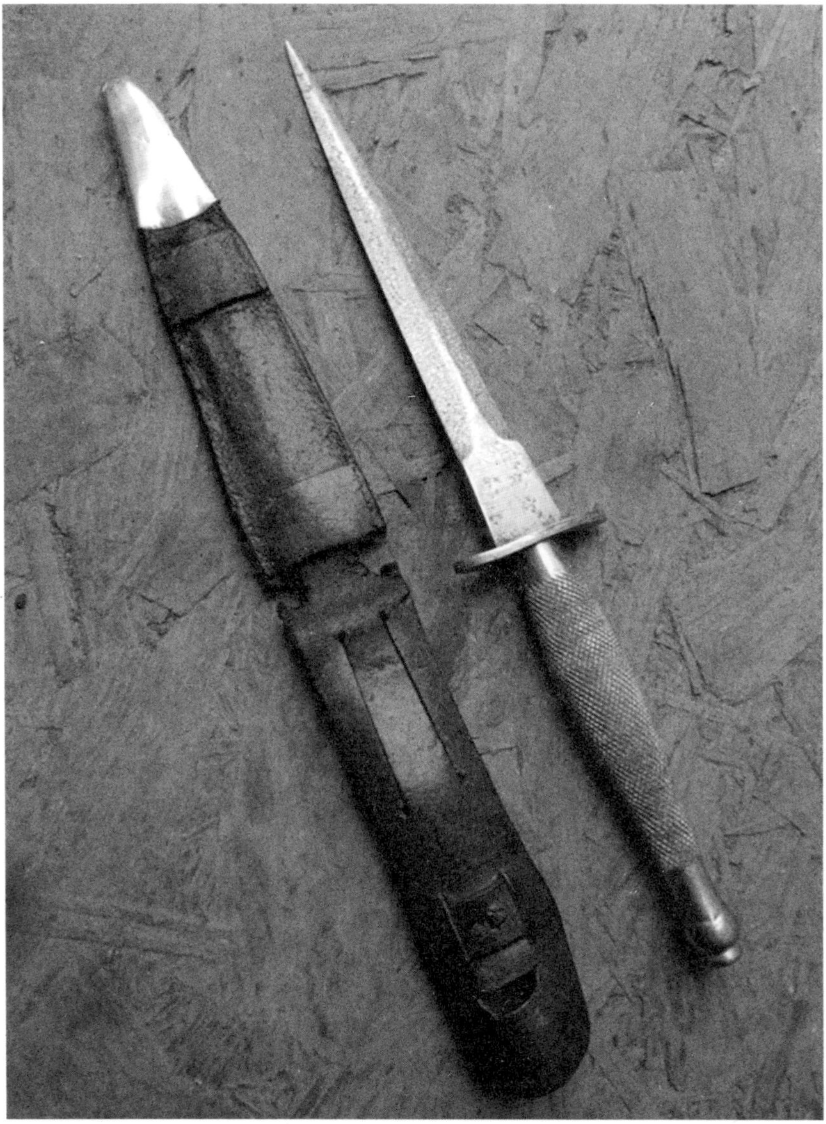

Fairbairn-Sykes fighting knife, 1st Pattern. (Author)

After his stint on the Peninsula, he moved into policing work in Shanghai. As a regular beat, this sprawling, cosmopolitan metropolis made Al Capone's Chicago look like Emmerdale; organised crime wasn't just rife, it was an institution. Fairbairn honed his knowledge of martial arts (the first European black belt in Judo), developing his own bespoke hybrid version. He liked blades just as well; he is said to have survived some 600 fights, building up a mass of scar tissue from defensive wounds but leaving many deceased opponents.

With his colleague Eric Sykes (not the comedian), he both trained his fellow officers and developed their joint unique weapon, the famous F-S fighting knife, A modern stiletto that every renaissance assassin would have coveted, it has a 6.5-inch tapering blade leading to a needle-sharp point. It's not a utility tool, it is designed for killing, for thrusting into exposed areas, sliding between unsuspecting ribs and generally disposing of unwanted sentries and other inconveniences. The tapered handle compliments the heft and facilitates a number of grips.

By 1940, Fairbairn was 55, but Hitler had plans and so did the Secret Service, recruiting ex-police to train emerging Special Forces, including British Commandos and US Rangers. Fairbairn never sugar-coated the pill. His creed was to do unto others as they'd like to do unto you, just do it first and better. Today's SAS still carry F-S knives. Sykes was an equally formidable character, reputed to be able to draw, cock and shoot a pistol from a shoulder holster rig in under a second.

Fairbairn wrote that some considered his methods of fighting to be ungentlemanly, but then Hitler wasn't much of a gentleman. Those Norwegians who had escaped the jackboot polluting their country and who joined *Kompani Linge* would have no cause to quail at his methods. Under the tutelage of Fairbairn and others like him, they would become masters of ungentlemanly warfare. Their road was long, hard, dangerous, and lined with repeated frustrations, but they would get the chance to 'do their bit'. And their bit would be splendid.

SOE's prime tasks throughout the occupied nations were threefold: (a) to circulate propaganda, undermining the oppressor and weakening his and collaborators' morale, (b) to build up local resistance forces to support allied intervention and (c) train those forces up as efficient guerrillas/saboteurs capable of inflicting steady but relentless attrition.

Norway at the outset appeared something of an enigma. In June 1940, just after the fall, Section D prepared a paper looking at tactical options which considered how best to 'stimulate the civilian population to undertake guerrilla warfare against the German occupying troops, and ultimately to extend guerrilla tactics to open rebellion'.[1] Rather optimistically, these Baker Street planners were hoping for a big push that autumn and when that clearly couldn't happen, a similar effort in the spring of 1941. It was vaguely envisaged that weapons could be brought in by fishing smack and submarine, grabbed by a populace just raring to have a crack at Fritz, who would never know what hit him.[2]

Rations would also be supplied to this shadow army of fighters, not just salty fishermen but urban professionals. 'Attacks by guerrillas during the dark, cold winter months would have a very strong effect on the already bad morale of the German garrisons.'[3] SOE fantasised that the feckless Hun was on drugs and that he'd be disorientated by being shot at from out of a snowdrift – a new, if rather unlikely, variant of a hedgerow war. Daft as this concept was, it brough forth the seed of an idea that, say, 30 or 40 young Norwegians trained in the UK could be infiltrated back into their homeland to build up resistance cells. Whilst the Government in Exile would be included in preparatory planning, only because keeping it secret from them would be a practical impossibility. Without any basis in reality Baker Street considered that 'The Norwegians as a people are notoriously ill-disciplined and are great talkers.'[4] Total tosh!

Sir Frank Nelson[5] reviewing these pious hopes injected a note of realism. Only a handful of Norwegian Commandos were busy training at this point, and he could see no immediate opportunity for any large-scale insurrection. This grand plan gradually faded into obscurity, though Hugh Dalton still held out hopes: 'All our reports go to show that, following the winter, Norwegian civilian morale .... remains at the highest possible level.' He was convinced that the problem wasn't galvanising resisters, it was holding them back, the spirit of the Norsemen brimming. It was all wishful thinking, dictated by a need to impress Churchill. By the middle of 1941 this had been scaled down to preparing resistance groups to act either in the event of a major allied landing or a perceived collapse in German capacity and morale. Neither of these was likely at the time.

Martin Linge and other escapees were able to paint a far more realistic, not to say sobering picture. The Government in Exile appointed the soldier-turned-actor-turned soldier as their liaison officer with SOE and he proved to be an excellent choice. Norwegian Independent Company Number One ('NIC1') was effectively his brainchild and fittingly was named for him after his death in action. All of those men who were to participate in *Grouse/Swallow* and *Gunnerside* would pass through the ranks of *Kompani Linge*.

Initially, these exiles remained in civvies but during July 1941 they became a discrete unit within the ranks of the Royal Norwegian Army – coming in at a rate of a score or so every month until building up to around 250 by mid-1943. Their training under the aegis of SOE began at Brackenbury, Hertfordshire (STS 17), then up to Fawley Court, Henley-on-Thames (STS 41) and then after November 1941, north to the next best wilderness to the Hardanger Plateau, the Cairngorms near Aviemore (STS 26).

SOE itself was a backstairs bastard born out of necessity, something of a mishmash from the start. The organisation was created from three existing but divergent departments. Spurred by the Anschluss, proof of Hitler's evil intentions, the Foreign office set up its own desk – Department EH.[6] Not to be outdone, Secret Intelligence Service formed Section 'D' ('D' apparently for destruction!) headed by Major Lawrence Grand. Finally, the War Office, not wanting to be left behind, set up or rather expanded its research wing, Department GS (R) to embrace irregular warfare, led by Major J.C. Holland.[7] Needless to say, there was any amount of sloppy overlap and inevitable internecine bickering.

As the war gathered momentum before sliding into apparent stalemate, Section D had a stab at sabotage in an effort to mine the strategic bottleneck of the Iron Gate along the Danube. Section MI (R) produced beginner's guides to guerrilla warfare but also kick-started the idea of independent émigré companies as spearheads for taking the fight back to the enemy.[8] Still, Britain wasn't quite setting occupied Europe on fire just yet.

Winston wanted more – a lot more. On 1 July the war cabinet decided there should be a single body to undertake infiltration and sabotage right across Hitler's swollen fiefdom. On the 22nd of that month Sir Hugh Dalton,[9] Minister of Economic Warfare, was named as political head and instructed to 'set Europe ablaze', a typically grand Churchillian flourish. Ironically, Dalton settled on the IRA as the most effective template. Nelson was now appointed to run the whole show, but Major Holland's previous deputy Colin Gubbins was brought back into the fold. He had been heading up the Home Guard Auxiliers (stay-behind saboteurs, very much not Dad's Army) but became SOE Ops Director. That ubiquitous department, dubbed 'The Churchill Toyshop', responsible for the eclectic array of weapons of minor destruction maintained its independence – exploding rats a house speciality.

In February 1942, Dalton was 'stellenbosched', as might have been said in the previous war, (The term probably originated during the Second Boer War and refers to a transfer to some forgotten outpost for those officers who'd offended the wrong superior – the French equivalent is *degommé* – literally to become unstuck.)

Some felt he was too left leaning and too prone to using phone taps for eavesdropping on his Labour colleagues. Lord Selborne took over as Nelson retired on health grounds. Sir Charles Hambro, a successful and influential city banker, picked up the reins. Hambro was experienced and vastly energetic, close friend to the PM with a sound war record (he had won the MC), behind him. Hambro, who never seemed to let his commercial interests affect his war work, stayed in office during the period when *Grouse/Swallow*, *Freshman* and *Gunnerside* were active. Gubbins finally took over as director in September 1943. During the war SOE was never a constant, it couldn't be, as strategic demands kept shifting and the theatre of operations expanded.

In very general terms, the organisation was tripartite: propaganda, operations and research. Inter-departmental and political rivalries never disappeared but here we are primarily concerned with Operations, which ran missions into enemy (and often neutral) countries. Some like France had no fewer than six sub-sections operating, but each occupied nation with its own dedicated section(s) was compartmentalised. SOE itself was run by a committee comprised of a dozen and more departmental or section chiefs. Most were from the armed services, but civil servants, lawyers and business types were also represented. As any reader of spy fiction might divine, recruitment was generally informal and based on the 'old boy' network; public school alumni and Oxbridge graduates featured heavily.

SOE spread its tentacles as the war itself spread and operations proliferated across the globe. Whilst the mission statement was to sabotage and subvert, many direct missions were mounted. Ultimately, the Norwegian effort would be the most successful, a brilliant surgical strike with no casualties either amongst the Commandos themselves or the civilian population. Others, such as the assassination of the odious Reinhard Heydrich in Prague, *Anthropoid,* resulted in the loss of the whole team and terrible reprisals. Relations with various governments in exile could be consequently fraught, not to mentions friction with the FO and SIS, who preferred resisters to act as intelligence gatherers rather than taking on the Axis direct – this led to disasters, such as Vercors in France during 1944.[10]

At least at the outset, SOE worked well with Combined Operations HQ. Mountbatten saw himself as something of a buccaneer and the new weapons and tools developed by Gubbins' teams proved very useful in Commando operations. Inevitably perhaps, the Admiralty didn't take too kindly to amateurs messing in naval matters and Arthur 'Bomber' Harris was grudging in his loan of aircraft. Stewart Menzies, heading up SIS, disliked his rival's swashbuckling, especially if, as did occur, SOE backed communist partisans on the grounds they were more active. On political grounds, SIS was none too keen to encourage the forces of the far left.

It was at the end of October 1940 that SOE moved into its wartime premises in 64 Baker Street, with the inevitable consequence of its members becoming labelled 'The Baker Street Irregulars'. Pretty soon the organisation had taken on most of the west side and reference to SOE was often just 'Baker Street'. Like the later fictional 'Universal Exports' of Bond fame, SOE didn't exist in terms of the regular postal service or phone directory. Its reach extended far, far beyond the streets of the capital and out into a whole network of sub-units, research stations and training centres, invariably set up in requisitioned country houses. The wags soon decided SOE stood for 'stately 'omes of England'.

Special training schools spread out from places such as Wanborough Manor near Guildford to the altogether more rugged Scottish Highlands (see below). Arisaig is a beautiful region with marvellous beaches, but SOE agents went there to learn lethal skills under virtuosi like Fairbairn and Sykes, (more on this later). Those who got through this stage went on to be inducted into the realms of spy craft at SOE's finishing schools around Beaulieu in Hampshire. Having 'passed out' they could undertake parachute training at STS 51 & 51A near Altrincham in Cheshire then, under the aegis of No. 1 RAF Parachute Training School at Ringway Aerodrome (now Manchester Airport).

All those young Norwegians destined to take part in the raid on Vemork would 'graduate' via this process. What was remarkable and far thinking for its time was the organisation's willingness to select purely on merit. SOE recruited many female agents, known homosexuals, those with past criminal records, communists, and anti-British nationalists – a huge contrast to their Nazi foes who were always guided by their own limitless bigotry.

Radio was, of course vital. SOE had no other effective means of tracking, supporting or instructing its agents in the field. Every mission had at least one skilled radio operator. Once Team *Grouse* was dropped onto the Hardanger Plateau, the overall mission could never have proceeded without the team's dogged and resilient use of comms. In the early days, radio traffic went through

SIS's stations at Bletchley. SOE went live on its own account on 1 June 1942, based at Grendon Underwood in Bucks and adjacent Poundon. Teleprinters communicated directly with Baker Street.

Virtually all of SOE's early radio sets were hand-me-downs from SIS. These were big, heavy, cumbersome, and notoriously underpowered. Better kit was provided by exiled Polish forces. Finally, as the organisation came of age, they manufactured their own gear. The Paraset[11] weighed only 9 lbs (4 kilograms) and could easily be concealed in a small case. That was more like it, and a powerful set, the B2, heavier at 32 lbs (15 kilograms), had an effective range of 500 miles (800 kilometres). With time and experience (and losses), radio operating procedures became more sophisticated and secure – the radioman or 'pianist' was an essential player in any operator team. Like their first radios, SOE's ciphers were another throwaway from SIS, though the service was blessed in having Sir Leo Marks as its chief cryptographer.[12]

One area in which SOE proved singularly inventive was the creation of that whole range of ingenious explosive devices designed to abbreviate the lives of as many Nazis as possible. This wasn't in the least bit gentlemanly – it was a form of state-controlled and directed terrorism, and terror has always been effective. SOE engineered a whole demolition derby of limpet mines, shaped charges, and time fuses, all of which were lapped up by Commandos and even SIS. The time pencil, designed by Commander A.J.G. Langley, enabled the saboteur to make good an escape before his targeted balloon went up. It was much easier to use than a lighted fuse or electrical detonator. Once an internal vial of acid was crushed, this corroded a retaining wire. Initially temperamental in terms of the vital time factor, an improved model, the L-Delay, less affected by temperatures, proved significantly more efficient.

The service developed the use of plastic explosives, a lethal brand of plasticine that could be cut, shaped, and fitted with impunity, easy and safe to store, ideal for saboteurs and resisters. The IED had just come of age. There were special lubricants cut with grinding dust that could damage any vehicle or locomotive working parts, the now famous exploding rat, land mines masquerading as animal dung, exploding pens and a pistol concealed in a smoker's pipe –another smoking hazard. If none of these worked, an agent could fall back on earlier technology and wield a sledgehammer to crack machine mountings. We can see where Ian Fleming got the inspiration for 'Q'.[13]

Transport was a permanent headache, SOE did experiment with its own submarines and obviously, it was always possible (as would be the case with *Grouse/Swallow/Gunnerside*) that agents could get over frontiers on foot or by hitching a ride. But aside from the obvious risks, violating neutrality could have diplomatic consequences. By air was the best means. That meant the RAF, but many senior airmen, such as Air Vice Marshal Portal, viewed their undercover colleagues as little better than gangsters and assassins, whose silly games jeopardised aircrew and disrupted the RAF's primary roles. Arthur 'Bomber' Harris was no more enamoured and intensely disliked the notion of risking his aircraft and their precious crews to accommodate mere adventurers.

To the disgruntlement of many, SOE were able to get hold of a brace of Whitley bombers. The Armstong-Whitworth Whitley was a twin-engined medium bomber designed to meet Air Ministry standards in the mid-1930s. Though an effective workhorse, it was obsolete by 1942 when replaced for front-line service by four-engined 'heavies'. It was re-designated to a secondary role in coastal defence, training and as a glider tug – in which role it was very much underpowered.

These seeds germinated into 138 Squadron (flying arms and supplies) joined from February 1942 by 161 Squadron (ferrying agents). Both units were initially based at RAF Tempsford in Bedfordshire, though 161 Squadron frequently few from RAF Tangmere, further east, which cut down on flying time. Tempsford was one of Harris's rejects, ground being too wet for heavy bombers, so the place stayed looking like your average farm layout. SOE used Tangmere Cottage by the main entrance. Agents were ferried in from local hotels to 'Gibraltar

Farm' – their final briefing and tooling-up point before the moment of departure. Then that defining voyage into the dark, all with the certain knowledge they might never be returning and with their cyanide capsules a far better alternative to capture and inevitable torture.

Gubbins' general belief was that agents should know how to handle standard issue weapons, Colt .45 and .32 automatic pistols, .45 Thompson sub-machine gun (the famous Tommy-Gun and 'Typewriter' of pre-war gangster movies), the 9mm Sten, which, for all its crudeness and limitations was, in the right hands and if kept clean and well-oiled, an effective SMG at close quarters.[14] 'Cheap as chips' to produce and able to take apart into sections which made for easier concealment. Some bespoke weapons, such as the De Lisle carbine[15] and the ingenious Welrod pistol[16] were added to the secret arsenal.

An SOE agent needed to know how to kill, and not necessarily at a distance, quite the reverse. He or she had to be able, quite dispassionately, to dispatch a target or adversary at close quarters. Hesitation was apt to have severe consequences. Before they began their SOE training, these young Norwegians from *Kompani Linge* hadn't killed at close quarters, even though most had fought in defence of their country before escaping to Britain. Shooting an enemy soldier over a distance as he is shooting at you is essentially a question of survival and you pull your trigger in the red mist of battle. By comparison that's easy.

Willie Fairbairn well understood the difference:

> There are very few men who would not fire back if they were attacked by a man with a gun … But suggest that they retaliate with a knife, or with any of the follow up methods explained in this manual, and the majority would shrink from using such uncivilized or un-British methods … Killing with the bare hands at close quarters savours too much of pure savagery for most people.

Fairbairn proclaims that his students will be fighting not just for their lives but that of their nation, so 'squeamish scruples are out of place.'[17] Training was thus of prime importance and aspiring agents had much to learn.

M.R.D Foot likens SOE training to 'a set of sieves, each one with a closer mesh than the one before'.[18] Gubbins knew how vital this was. Motivation wasn't enough, hatred of the Nazis wouldn't do on its own. These young Norwegian men who were about to undertake one of the most important missions of the war, had to be fully ready in every sense, one weak link would be fatal. Initially, hopefuls were all sent to one of SOE's country house locations for a couple of weeks' induction. This wasn't Agatha Christie, those halcyon days of pre-war house parties were only a memory, it was far more Cold Comfort Farm.

Being physically fit was obviously an essential, the levels of endurance *Grouse* and *Gunnerside* agents would display are astonishing, their stoicism and sheer grit unparalleled. They were born with spirit and love of their beleaguered country drove them on, but it was SOE who honed and sharpened them. The bar in every STS was well used but there was no such thing as a free drink. Handlers would be watching to see how agents behaved in the face of temptation and if they could hold their liquor. Cross country runs, fitness training, 'beasting' as it might be called today, alternated with essential small arms drill. Instructors were all vastly experienced, as were team leaders such as Roger de Wesselow, a Guards officer who still led from the front despite being more than twice the age of most of his students.[19]

Those who survived this starter pack were next despatched to the highlands for more intensive Commando style training, as mentioned previously, in the hauntingly beautiful region of Arisaig on the west coast. Though steeped in Jacobite myth, few recruits would get to savour any sense of romance. It was tough and it was dangerous. Gubbins had managed to get the whole area cordoned off as a 'no-go' zone. I spent many of my school summer holidays there in the 1960s and fell hopelessly in love with the place, but that was in summer, and I wasn't on a 20-mile tab in full kit during the depth of a near arctic winter, in wet wool on little or no sleep.

Firearms training was more intensive here, and operatives were expected to familiarise themselves with current small arms from all combatant nations: revolvers, pistols, rifles, SMGs and light machine guns, especially the BREN. These drills included, besides loading and firing, stripping, re-assembly cleaning and maintenance. In the field you'd not have an armourer in attendance. Self-reliance was paramount.

It was here they would meet Fairbairn and Sykes, those magi of nasty warfare. Both were well into middle age and had a vast cabinet of violent experiences to draw upon. This dynamic duo, as Fairbairn states in his book, were there to turn essentially ordinary young men and women into ruthless and instinctive killers. Back in Shakespeare's day fight master George Silver had similarly educated his students in streetwise swordplay, none of Tybalt's Italianate flourishes, just about winning the duel and employing whatever tactics might finish the job fastest. These didn't have to be pretty. The objective was a dead or disabled opponent with least personal inconvenience. Agents in the field were set far apart from regular forces, including Commandos, in battle. They'd be very likely on their own, deep in enemy territory with nowhere to retreat to. The enemy in front of you would have to be killed quickly and silently and your actions must be immediate.

Both former detectives were also crack shots, authors of the now famous 'double-tap' – 'one through the heart and one through the head, now the bastard's doubly dead.' The process was cumulative, agents were turned into killers, not just in the sense of their ability to execute lethal moves but in their confidence and assurance, 'a sense of physical power and superiority that few men ever acquire', as one trainee described it. Fairbairn and Sykes would go on to induct recruits in Canada together with personnel from Bill Donovan's nascent OSS. Their methods would influence Special Forces training from then on – as they continue to do.

Candidates were rigorously and regularly assessed; this was the world's most exclusive finishing school. The CO of the unit would consult both with instructors and 'conducting officers' – drawn from recruits' home nation and, ideally, already experienced agents. This wasn't just about caring and sharing, as SOE had to be completely sure of its personnel. Not only was the organisation investing heavily in their training and preparation but, in the field, a wrong choice could have catastrophic consequences. At the outset many volunteers (as brilliantly portrayed in Sebastian Faulks' 1999 novel *Charlotte Gray*), didn't quite know what they'd signed up for.

But for these Norwegians, it was different. They'd experienced Nazi terror at first hand, they'd seen their country stolen. They would endure whatever hardships they had to, so they could strike back. They were young, fit, and willing, in many senses perfect recruits. Most had lived their lives in or near the mountains. They knew what a winter storm felt like. Hatred sharpens an edge keen as any whetstone.

From small arms and unarmed combat training, recruits passed on to the noisily exciting world of explosives and carrying out sabotage demolitions. London, Midland and Scottish Railway was swayed to provide spare lengths of track plus rolling stock and locomotives, so they could be blown up! Away from satisfying big bangs, there was orienteering and map-reading, small unit tactical training, house clearance, ambushes. Grenades came in handy for dynamiting fish, enterprising perhaps but not likely to endear SOE to the local gamekeepers. Raiding each other's messes to secure additional drink was also a popular extension – alcohol, this being Scotland, was more severely rationed.

Norwegian recruits also trained at SOE's STS 26, located in the Cairngorms, a wild and rugged landscape, prone to severe conditions in wintertime (and not infrequently in summer). This formidable terrain with its steep hills, bare upland moor, gullies, twisting streams and forest was as near to Norway as it gets in the UK. Temperatures could and did plummet, and even if not as low as in Norway, recruits felt the raw, dreach, dankness of Scottish rain more than amply compensated!

In the days of Queen Victoria's reign and the ensuing Edwardian twilight, wealthy shooting parties travelled by train here to follow their sport in comfort. Timber-clad shooting lodges proliferated, and they now proved very handy as training centres, only the customary bevy of servants wouldn't be coming. Drumintoul Lodge, two miles (3.2 kilometres) east of Aviemore was put up in 1878. It boasted four public rooms, a dozen family suites and eight garrets suitable for servants. Naturally, there was a tennis court.

A very similar structure, Glenmore Lodge, stands four miles further east of Aviemore and was chosen for a royal visit whilst Norwegian recruits were training there. This description accompanied a picture of King Haakon:

> Picture taken at STS 26, Glenmore Lodge, during a demonstration by *Kompani Linge* for the Norwegian King Haakon 7th, and Crown Prince Olav, Norwegian Minister of Defence during World War II and later King of Norway. The 'dead' German soldier between the railroad tracks had been rendered harmless and an explosive charge placed. The saboteur had crawled up on the hill on the left from a position just behind where King Haakon is standing ... then jumped off the hill at the German sentry and killed him with the Japanese *hadaka-jime* stranglehold.

Drumintoul and another centre, Forest Lodge (1880) were sited in woodland, with lochans nearby. Forest Lodge was the biggest of its kind in Scotland. Being as a rule timber-clad, more than a few succumbed to fire over the years, but shades of their Edwardian heyday clung on. Quite what the tweed-clad 'guns' would have made of the new tenants is uncertain.

At Arisaig/Cairngorms Group 'A' schools, about a third of the candidates might chuck it in or be told they'd not made the grade, 'RTU'd', but there was no opprobrium in this. Failure wasn't a badge of shame, it just reinforced how high the bar was set. Those who did make it through moved on into Group 'B'. More country houses, but back down south again around Beaulieu. By now the students were allowed to understand the true nature of their mission, that they'd be infiltrated into enemy territory. They would learn about who and what they were up against, how occupation worked, the structure of Axis security units, how they were deployed and the role of auxiliary police services, the extent and nature of local collaboration. After the almost surreal world of mountain and dale, this was a swift introduction to the rim of the abyss.

Agents who had made it back gave lectures on how to deal with routine identity checks. And there would be interrogations. Agents would be dragged from their beds and subjected to a very realistic going over, short of physical abuse. This smacks of crude theatre, but in fact such training did serve to prepare agents for the first shock of capture. More pervasively, recruits began to grasp and embrace the nature of that double life they'd be leading once infiltrated behind enemy lines.

They were to be actors as well as players. Their cover, or 'legend', had to become an essential part of who they were, so that like master thespians, they could inhabit the role they were playing. This was serious drama; their lives could very well depend upon it. The role-play was immersive; language, dress, down to underwear, smoking habits, ordering food or *un café*. Travelling by train or bus. For the Norwegians this at least would be easier. They'd be operating on their home turf, speaking their own tongue. Still, a single slip could be fatal. Complacency was never an option.

Arisaig taught hopefuls how to be competently and instinctively aggressive, Beaulieu taught them how to be self-effacing. Golden rule number one was to be neither seen nor heard, avoid anything that might arouse suspicion, don't be cocky! Despite this, naked aggression never disappeared from the syllabus, their continuing mantra dinned into them, almost as indoctrination, was to hate the enemy and be ready to do him harm wherever and whenever the chance arose. Hatred was their ally.

By way of fine tuning, agents were taught how to compile reports, elementary coding, recruitment of resisters and what post-war spy thrillers would call 'tradecraft'. How to check if you were being followed, how to shrug off a tail, dead letter drops. Once a candidate had convinced his/her instructors that he/she was ready, they would undergo a final empirical test. They were sent out in small teams to perform an allotted task, to pick up a hitherto unknown contact at a specified location and local police would be looking out for them. If they did get nicked, they had to rely on their cover story. SOE would be watching over its flock and would intervene before any of its foundlings ended up in the dock.

Once they had graduated from Beaulieu, their preparation still wasn't over. It was on to a series of specialised courses; parachute drops from Ringway, printing leaflets for propaganda, breaking and entering, picking locks (often taught by proper 'villains', recruited from those otherwise serving at His Majesty's pleasure). East of Oxford at Thame Park, SOE maintained its wireless school – a shortage of trained operators was a seriously limiting factor for SOE.

Some would study under legendary George Rheam who ran Station XVII from Brickendonbury Manor, located between Hertford and Hoddeston. Rheam was an industrial engineer by training who had worked for Metropolitan Vickers before the outbreak of war, recognised as the expert in electrical generators. What his mind could create, his hands could demolish, and perhaps his major contribution to *Grouse* and *Gunnerside* would be his development of tactics and bespoke explosive charges, aimed at knocking out industrial plant and gear. He would send his students out on mock raids against UK installations (without any warning to the intended target), so they got the taste for operations.

It is one of history's timely synergies that the Norwegians impressed him the most with their steady courage and their appetite for risk-taking. None of these teams could, at this stage, have been aware of just how valuable such tuition would prove. Rheam was dubbed 'The founder of modern industrial sabotage'.[21] His pupils, in the field, would know their way around industrial plant they were tasked to destroy and to do so with a minimum of explosives, just fixed in the right place, and inflicting sufficient damage to significantly hinder repair or replacement. Rheam is very much part of the Vemork story.

So was Martin Linge. The soldier/actor would be an inspiration to the younger men of Norwegian Independent Company No. 1 (latterly re-branded as *Kompani Linge* after the gallant captain's demise).[22] Operation *Claymore*, launched on 1 March 1941, was an early Commando strike. Nobody from *Kompani Linge* took part but 52 sailors from the Royal Norwegian Navy accompanied soldiers drawn from No. 3 and No. 4 Commando and they biffed targets located in the Lofoten Islands three days later. In total 11 factories were blitzed, and 800,000 gallons of precious fish oil were destroyed, 225 Germans with 60 Quislings were captured and over three hundred volunteers came back to the UK. Tools from the Enigma were also seized and proved of great interest to Bletchley Park. Cheekily, the Commandos sent a telegram direct to Hitler from the local post office, taunting the Fuhrer for the Germans' unpreparedness and feeble resistance.

On 27 September that year, 800 raiders, drawn from No. 2, 3 and 4 Commando with 101 Canoe Troop from 6 Commando joined by a dozen Norwegians under Linge, participated in Operation *Archery*, an attack on Vagsoy, beating up enemy quarters. Landed by the Navy, these raiders quickly discovered Intel had significantly underestimated the total of Axis defenders. Nonetheless, four fish-oil factories were wrecked with nearly a hundred captives going into the bag. The Commandos lost a score of dead and wounded and the Norwegians saw Linge fall, picked off by a random sniper. Back in Berlin, Adolf was not best pleased, persuaded that Norway might prove the back door in to his 'Fortress Europe' and deployed yet more troops, 30,000 in all, to beef up his garrison. That's almost two divisions effectively lost to the Axis order of battle.

SOE was to play a part in the later Operation *Musketoon* when on 15 September 1942 ten Commandos and two Norwegianslanded by sub – their objective to smash a hydroelectric plant at the head of the Glomfjord, which would significantly disrupt aluminium production at an adjacent factory. Germany desperately needed the product. Having tabbed over high ground for three days, the force struck on the night of the 19th. Significant damage was inflicted but resistance proved fierce. Half a dozen men were captured, including two wounded officers. As PoWs they were sent first to the legendary Colditz, but as such raiders fell foul of the infamous

'Commando Order' (*Kommandobefehl*), they were dragged to the even more notorious Sachsenhausen KZL. On 23 September, all captives were shot,[23] the first victims of Hitler's draconian edict. As an added cruelty, families, via the Red Cross, were told these men had escaped. All four of those who did get away at the time made it safely into Sweden.

If *Gunnerside* was by far the most important and successful SOE-led operation, it was almost immediately followed by the ambitious *Martin*. The objective here was to secure a foothold in the Tromso region and to disrupt German action against the vital North Atlantic Convoys. On 29 March 1943, the Shetland Bus infiltrated a team of eleven raiders, their mission to brass up seaplanes and take out the German air control tower located at Bardufoss. It didn't start too well when one of the Commandos contacted a local shopkeeper, disastrously an active Quisling who just happened to share his surname with a loyalist. The Germans were quick to react and captured all but one of the team, Norwegian diver Jan Sigurd Baalsrud, who made a run for it, minus one boot and with the enemy in close pursuit. He leapt into the dank chill of the harbour and swam towards a gully which offered a means of escape. The Germans thought so too, and he only got clear after a shootout during which he accounted for a brace of his adversaries.

Baalsrud's adventures were only just beginning. His only hope was to get across country and over the Swedish border. The journey turned into an epic of survival, in the teeth of snowstorms and avalanches. He cut off his own frostbitten toes without anaesthetic, yet struggled on. Despite his betrayal at the outset by collaborators, many loyal Norwegians risked their lives to help him and without them he'd never have made it. But make it he did and got back safe. His overall condition wasn't good, and he spent half a year in recuperation before returning to the Cairngorms as an instructor. Like those who carried out *Grouse*, *Swallow* and *Gunnerside*, Baalsrud became an icon, and rightly so, the human emblem of an unconquerable will to resist the jackboot and free his country.

Quite a lot has been written about tensions between SIS and SOE. These existed to be sure but probably impacted less on operations in Norway than might be assumed:

> There is a historical consensus that the Secret Intelligent Services (SIS) and Special Operations Executive (SOE) were, from the summer of 1940, locked into a conflict at the highest level, based on rivalry and jealousy, and which at times threatened the very existence of the SOE. It has also been claimed that the operational priority given to the SIS along Norway's west coast significantly restricted SOE operations into the country.
>
> However, examination of the SOE archives ... appear to show that at a sectional and operational level within Norway the relationship between the SOE and SIS was not fraught with the same difficulties ... The archives provide numerous examples of co-operation between the two organizations ... They exchanged intelligence, shared clandestine radio stations, and used the same fishing boats to transport agents to Norway; several SIS agents moved on to work for the SOE ... It was a relationship more often built on co-operation than conflict.[24]

Weather on the moonlit night of 28 March 1942 was fair and Einar Skinnarland was uncomfortable inside the fat-bottomed belly of a Whitley Bomber headed out over the North Sea bound for his native Norway, from which he had only escaped a couple of weeks before. He probably didn't quite get it at this stage, but he was the sharp point, the *schwerpunkt*, of what would become *Grouse* and then *Gunnerside*. For this young man, it was a testing time, with the diabolical racket of the Rolls Royce Merlins, the queasily shaking motion of the bare fuselage and pit of fear in his stomach, thinking of the drop ahead.

Just those two weeks previously, Skinnarland had been on the operating table in a Norwegian hospital at Kristiansand, awaiting surgery on his left knee. It didn't look promising, damaged in a fall the joint was badly swollen. He refused anaesthetics, injury or not he had urgent plans for getting out of Norway. So this was going to hurt, and it did. With commendable sangfroid, the patient endured without a squeak and next refused the necessary bed rest, a very tight dressing and the aid of a walking stick would have to do. Hobbling into the dock area he was able to gain a brief respite in the back room behind a chandler's sales shop. Rest wasn't really an option as the agony from his battered knee was blinding. In his pack, like Billy the Kid, he carried a pair of revolvers. He carefully checked, cleaned, oiled and loaded both. Any minute now the 620-ton Norwegian freighter *Galtesund*[25] was due in harbour. Einar Skinnarland had plans for this vessel.

Einar was a native of Telemark, the eighth son of Hans who was the dam keeper at Lake Mos, high above and west of Rjukan. Whilst the plant at Vemork was state-of-the-art, the Skinnarlands were timeless, their timber-planked eyrie reachable only by skis during the long hard months of pre-war winters. Self-reliance was not simply a lifestyle choice and young Einar swiftly became an expert skier. Acceptance of hardship, privation, patience, and endurance were in his DNA. He moved on skis, saw friends on skis, skied to school, his skis were virtually limb extensions. Still, he was an apt and assiduous student. High marks gained him a place in college to study engineering. He also trained for war (the one Norwegians never expected to arrive), at the engineering corps academy in Oslo, marked out as officer material.

'You may not be interested in war, but war is interested in you' is an aphorism almost certainly misattributed to Leon Trotsky. In April 1940 Norway discovered it to be entirely correct. Despots pay no heed to protestations of neutrality. Einar, like everyone else, played his part but to no real effect. Both of his brothers Torstein and Olav, more than did their bit, scrapping with the invaders around Rjukan and straining every last sinew of organised defiance.

Einar took off his uniform and returned home, but remoteness was no refuge. Already the family hotel by the dam had been requisitioned, the hated *feldgrau* every day in the family's faces. He went to work at Vemork as a construction engineer. He found a girlfriend and as it happened, his pal Olav Skogen was going out with her sister. Pre-war they'd just have been a happy quartet of young people, at home on skis or bicycles. Both lads were resisters, active in the fledgling Milorg, recruiting, collecting and stashing guns, working on illicit broadsheets. Their war was very far from over, it was only just beginning. Resistance might be swelling but so was surveillance. A local lad picked up by Nazi security dropped Skinnarland right in it. They couldn't find any weapons but did pick up a radio – these were illegal. Einar spent over a week in the slammer but was released without charge. He remained undeterred.

In fact, his ardour increased, chafing at the bonds of occupation. Knowing also that next time he wouldn't be so lucky, he asked Milorg to get him over to Britain to train as a clandestine radio operator. Under the pretext he was setting forth on an extended hunting expedition, he packed and left the dam. In fact, he was headed for the coast for a small port west of Kristiansand where he'd be picked up by boat. Instead, he bashed his knee and so ended up beneath the surgeon's scalpel.

Before sustaining his injury, he'd RV'd with a pair of other resisters, one of whom, 25-year-old Odd Starheim, was a mariner himself with a master's ticket. Earlier in the occupation, Starheim had escaped to Scotland and joined Linge as a very early recruit. Returning, he'd been hyper-active in resistance circles, focused on German shipping. Inevitably, he had the Gestapo on his tail and urgently needed to make that same sea-crossing. At this stage their evacuation by the RN was scuppered by bad weather. They'd need to find their own taxi and Starheim, together with his number 2 Andreas Fasting, had his eye on the *Galtesund*.

A mere 48 hours after surgery, Einar Skinnarland, stick in hand, boarded the target vessel as a foot passenger for an overnight coastal run from Kristiansand to the small port of Flekkefjord. No Germans, only one other passenger, a hefty crew of 22 with an experienced older skipper. As a nice touch, Einar sported an NS pin in his suit lapel. He also had both revolvers in his baggage. When e prepared for the return journey, Starheim and Fasting would also be aboard. Their intentions were wholly ungentlemanly.

At 17.00 hours local time, the ageing workhorse weighed anchor, propellers churning the placid spring waters. Just a routine trip. The crew would have been less relaxed if they had seen Einar leave his cabin, both guns tucked into his belt gunfighter-style, and with several lengths of rope. At 18.20 they were ready for the holdup, though Starheim had to explain to the 2nd mate why he'd boarded without a ticket. The experienced resister glibly lied his way out of any embarrassment and the matelot departed satisfied. Minutes later, the pair burst into the saloon like Butch and Sundance, seizing the astonished (and outraged) master with his single passenger. Tying their captives up, they headed for the bridge while Fasting secured the engine room. In a swift and bloodless coup, *Galtesund* was theirs. All that now remained was to get clear of the Oslo Fjord, across the wild North Sea to Aberdeen whilst avoiding the unfriendly attentions of the Luftwaffe. Piece of cake.

First, they had to keep to their advertised course to avoid suspicious glances from Axis coastal fortifications. Only when darkness fell could they set course for Scotland. Meanwhile, their comrades back in Norway had sent a wireless message informing London of the capture and their need for long-range fighter cover. Starheim and Skinnarland held the bridge through those precious hours of darkness, fighting off exhaustion with caffeine. Fog blanketed the sea next morning, the calm broken by the rumble of a plane's engines.

Ours or theirs? Turned out it was one of theirs, but the misty shroud hid them. It was around 1400 that the fog lifted. Visibility was now good, and they were terribly exposed. Another aircraft came roaring in, but this time it was one of ours, a mother hen come to protect her chick. Next morning, they sighted Aberdeen, guided safely in by a local armed trawler. Salvation. Einar almost immediately found himself enjoying a less stressful mode of travel, the 'Aberdonian' – rattling through the early spring night, all the way south to King's Cross.

Leif Tronstad, alerted by John Wilson, was waiting for him inside an average block of flats called Chiltern Court, located not far from SOE's main lair on Baker Street and blandly badged as the 'Inter-Services Research Bureau'. Wilson wanted Trondheim's take on the new arrival. Starheim had given him a glowing testimonial, but it was mention of Vemork that pricked up ears. Neither the professor nor the escapee engineer had met before. The SOE pair had two questions for Einar: would he already be missed and if not, would he mind awfully being parachuted back onto the Hardanger Plateau to keep an eye on the enemy's doings at Vemork? No, he couldn't be told why and yes, this would be very soon. Naturally, he was up for it.

His mission would be to focus on the manufacture of heavy water: how interested did the Germans seem, how well defended was the place, which of the staff might be loyalists? Skinnarland could see what his new masters were interested in, but they couldn't possibly tell him why. The world at large knew virtually nothing about nuclear fission and even less about super-bombs.

Bombing Vemork had so far been the preferred tactical option. 2nd Lieutenant Jens Anton Poulsson from *Kompani Linge*, a local fellow himself, with a small team, was more than halfway briefed for insertion onto the plateau as pathfinders for the RAF. Tronstad was extremely wary of such a sledgehammer approach. Bombing was far from precise, the target was extremely difficult to hit, the vital facility was located in a basement and if not bomb-proof, pretty close. The damage to Norwegian domestic economic infrastructure was potentially immense, not to mention the inevitable loss of civilian lives.

Now he had a very valuable new asset. Poulsson's squad could be sent in later, perhaps during April and identify a raft of targets which could then be surgically sabotaged, potentially far more damaging than waves of bombers whilst avoiding economic and civilian losses. Einar could be both inside man – after all, why should he not just simply return to work after his extended leave – and radio operator, the life-giving breath of clandestine operations. Welcome to SOE. Tronstad was very impressed with this new paladin who'd accepted his challenging brief without demur, to the extent they went out together for a drink. Both men probably needed one.

Neither Tronstad nor Wilson would have any cause to regret their choice. If Einar Skinnarland was feeling uneasy lodged in the belly of the Whitley barely a fortnight later as it droned over the rock and scree-ridden plateau of the Hardanger – and not just from the motion – who could have blamed him? He'd had a crash course in the several levels of SOE initiation, two weeks instead of the usual ten. Now with three stripes on a wool serge uniform inside his white jumpsuit, beside packed containers of assorted gear and weaponry, including a score of F-S fighting knives – and a souvenir spoon showing the Houses of Parliament, a gift for his mother – he was on his way. The moment of decision, point of no return, was rapidly approaching. He'd done well, impressing his instructors on his radio course at STS 32 and in parachute training at Ringway, three days instead of seven!

His jump instructor had offered the advice that when jumping from an aircraft, one was bound to descend and reach land, gravity would take charge of that. The manner of that landing was up to the parachutist! Whilst entirely logical, this wasn't necessarily helpful. Jumping by day in clear light, with England's green and pleasant pastures spread below, was one thing. But flinging yourself out, relying on the static line, into the howling black and freezing night, down onto a primeval deck of rock, or a frozen lake or sparse woodland, was really something else.

11.45 and the dispatcher gives him the nod. He is very alone. Out go the containers, slipping away through a circular hatch in the plane's innards, whoosh and they're gone, out into the raging pit of blackness. Einar edges toward the rim. Beneath his dangling legs, nothing but the night. Are they over the correct drop zone? He thinks not, or is it just his nerves, like a new-born reluctant to leave the warmth and safety of the womb? For 20 minutes this game continues, he's convinced himself they're in the wrong place but still, this big beast can't circle for ever, they're at the furthest extent of operational range. Sergeant Fox, the dispatcher, yells that fuel is now critical and if he doesn't jump, they will have to abort. He yells back that he's going to jump, and he does.

Norway rushes up to embrace him. He is going home – at 120 mph.

4

GROUSE

'Uncle, be careful. The town is full of Germans. Merry Christmas, Uncle.'

Heroes of Telemark

He with a short knife must try, try again.

Vapnfirðinga Saga, c.7

Second-Lieutenant Jens Anton Poulsson was very tired, he was hungry, verging on starved, his great reserves of endurance and stamina strung out to the limit. The other three men of *Grouse* team were in no better state. It was 23 December 1942, and the Hardanger Plateau was no place for ravenous, sickening and exhausted men. Poulsson had, for day after day, skied over this frozen, primeval place searching for reindeer – fresh meat was their only hope of salvation, it fuelled their desperate dreams.

He had the means, his Krag rifle was more than adequate and he'd a decent supply of ammo. He was a crack shot but really, really needed something to shoot at. The physical and psychological burden was immense, constantly being ratcheted up as his strength drained. Just existing in such a winter wasteland, even when properly fed, was a challenge. Right now, it felt like a losing battle.

Christmas was two days away but there'd be no turkeys this year. The *Freshman* debacle (see next chapter) hung over them like a shroud. They were now all there was. Of course, they didn't know the full import of their mission, they kept going on faith but even that has its limits. Once again, Poulsson focused his binoculars and scanned the horizon, a study in white, nothing moved. But suddenly something, several somethings were indeed moving. A jolt of residual energy spurted though his jaded system. Reindeer. Deliverance, if he could get close enough and kill one.

Adrenalin and a hunter's instinct kicked in. He had to work out his next moves with infinite care or the timorous beasts, their well-tuned antennae and impeccable noses, ever alert to risk, would spur them to bolt, hope pelting away into an impossible distance. He dared not fail. Keeping downwind, he crept slowly, carefully but surely towards them. At this point, the entire success or failure of their mission hung on Poulsson's skill and luck as predator. Reindeer have no interest in heavy water, but they could contribute mightily to its preservation or destruction at this very hour.

At the base of a frozen valley, a dark, fairy tale landscape, the reindeer paused to graze on what little there was. Poulsson's world was entirely focused on theirs, if they saw him, they'd stampede. If they smelled his unwashed human stink, they'd be off. He was at the mercy of a fickle arctic breeze. He needed his targets to work their way upwards into a jagged anarchy of crags, where the rock would provide him with adequate cover. But the reindeer weren't in any hurry. They had all winter. Daylight, too, was rationed and what feeble warmth there was began to ebb. He could feel the hungry fingers of incipient frostbite begin to search for his exposed face.

A chance. Two young adult males began to amble up the slope, still frustratingly out of range. Could he take the chance? Necessity moved him, and he began to work his way down towards them but slipped and fell. The bulls reacted, signalling their alarm to the rest of the herd. They all scattered, over the hill and far away. This could have broken a lesser man, his quest seemingly ruined as the sun dropped ever lower towards the horizon.

Happily, reindeer aren't very bright. Poulsson was and deduced they'd probably not veer off much further than the next valley.

With a supreme effort of will, propelling his tired body, Poulsson collected his gear and skis to set off once again in pursuit. He wasn't wrong; there they were and here was cover. He crawled forward with infinite nerve-shredding care and the herd was, at last, in range. The good news was that the regular cracking of ice in the frozen waste could sound like a gunshot. It might just be possible to drop one and the rest miss the message, leaving scope for a second kill.

He aimed the Krag (effective range 900 metres) and fired, the 6.5x55 round kicked against his shoulder. Nothing happened, the bullet must have missed, and the herd was already on the move. He banged off two more rounds to no better effect. It is impossible to conceive of Poulsson's agony. He'd had his chance to save them all, to secure the mission and he appeared to have failed completely. Command is a very lonely burden at times; success as they say has a thousand fathers, but failure is always an orphan. Poulsson knew just how that orphan felt.

But then he saw the blood trail and was instantly on his feet, tired heart pumping. The telltale slew of scarlet trailed over the rise where a young cow was stumbling. Two more shots completed the kill. Warm blood still spurted. Death meant life. Poulsson fished out his tin cup and filled it with bright gore. Drinking the blood of his prey spread fire through his whole being.

Poulsson and his team had been on the Vidda for over two months. A lot had occurred, hardship, isolation, tragedy and bitter disappointment. But at last, and even briefly, things were finally looking up.

Two shots fired in Sarajevo on 28 June 1914 propelled the European powers down a path leading to the cataclysm of all-out industrial war, a level of carnage not hitherto seen (other than,

Krag-Jorgenson rifle (MJL Militaria)

arguably, during the American Civil War of 1861–1865). Those three shots loosed by Poulsson on 23 December 1942, may not have resonated so volubly but if he'd not made that kill, the mission could very well have failed. Had the Nazis ever actually managed (however unlikely), to build a bomb, the world could have suffered even greater catastrophe. Let's hear it for the reindeer!

The beginning must have seemed a very long time ago, even if it was only months. Earlier in the autumn, Einar Skinnarland was already active after his dramatic drop onto home soil. He'd been able to radio back details of German troop strengths in the Vemork area; Rjukan held a company-sized garrison with a weak platoon up at his old family house by Mosvatn Dam. Both these detachments could reach the plant in around quarter of an hour. Thirty-plus enemy were stationed in a nearby school with twenty or so inside Vemork itself. Anti-aircraft defences had been increased; an undisclosed number of Flak guns set up with the added possibility of improvised minefields around the massive pipes that brought water down into the hydro-electric works. The Germans were being careful, guarding their precious 'jewels', as Von Falkenhorst described his supply of heavy water; but they were not yet on full alert. It was best for the moment whilst things stayed that way.

It was time for Combined Operations to deploy their *Grouse* team, a four-man unit, chosen and handled by SOE, intended essentially to pave the way for a full Commando-style assault on Vemork. Claus Helberg, in an extended post-war interview, told the story of how it began:

> In the summer of 1942, I was a member of the Linge Company's training camp in the mountains of Scotland. We were being trained for industrial and maritime sabotage and for training the Norwegian resistance forces. Professor (and Major) Leif Tronstad planned and organized the industrial sabotage. An excellent scientist, Tronstad was a member of

> the allied technical council. He made solid contributions to the Allies during the war, including the intelligence work behind the successful bomb attack against the German missile-testing site at Peenemunde. Tronstad was keenly interested in the heavy water production at Vemork. In 1934, as a chemical consultant, he had participated in planning the plant.
>
> One day in August 1942, the training camp in Scotland was informed that four men, including myself, were to report to Professor Tronstad in London. In London, our team leader received instructions for the assignment. To keep the information on the planned attack from leaking, only he was briefed on the real assignment. The rest of us believed that the action concerned training the resistance. By the end of August, we had rounded up the skis, boots, and other winter equipment we needed. Then we had to wait for a moonlit night and fair weather. For the next six months, the weather was our worst enemy, not the Germans. On our first infiltration flight, one of the plane's engines burned up, and we limped back to England. The next time we had to turn back because of heavy fog.[1]

These select individuals had been on standby for some time. Tronstad and Wilson first interviewed Jens Anton Poulsson, the designated team leader. Aged 24, never seen without his trusty pipe, lean, athletic with that 'hungry look' Shakespeare made so famous. The young second lieutenant was a native of Rjukan, where his family still lived, vulnerable to German reprisal and with the Germans, reprisals were always guaranteed. His handlers outlined the nature and scope of the mission but could not reveal its true importance. Poulsson and his men would be putting their lives on the line for a strategic purpose of which they would remain almost wholly unaware. This was asking a lot, but the Norwegian officer's only rejoinder after this initial briefing was, 'interesting'.

Knut Haukelid knew him well: 'The leader of the group was Jens Anton Poulsson, tall, gaunt and a good hillman. We had poached many a stag together and I knew he was in his element where shooting and fishing were concerned. He was, moreover, a sober fellow with sound judgement. He would manage all right where he was, for the Vidda was his own home country'[2]. Knut Haughland's skills as a radio operator would be vital, and he too was familiar with the Rjukan area. He'd be 2 I/C. Claus Helberg, a former schoolmate of Poulsson's, was third man. Even as a lad he'd been marked by his ability to both get into hot water and his uncanny knack of extracting himself without consequences. His Errol Flynn dash would be seriously tested in the months to come. Arne Kjelstrup might not have had the cavalier build of his comrades but the plumber from Oslo had great physical strength and grit and 'his inexhaustible humour might be coming in useful.'[3]

One man who would not be going out this time was Haukelid himself: 'But for an unfortunate accident, I should have gone with the four men of the Telemark Group who were already established on the Hardanger Vidda. We had been ready to start in the spring of 1942, but the plan was put off till the autumn on account of the light nights. During a field exercise I stumbled with a loaded pistol in my hand and put a bullet through the sole of my foot. I was recovering in hospital when *Grouse* party was formed.'[4] Haukelid's chance would come, and his role would be pivotal in the main mission's dramatic finale.

Skinnarland would be the vital link – he'd be advised of their arrival with an RV at a summer shieling located high in the Hardanger, an area happily shunned by the Germans, repelled by its primeval aspect. A message would come over via the BBC which would read 'this is the latest news from London,' rather than their habitual 'this is the news from London.' Once safely dropped, *Grouse* would contact Einar's brother Torstein at the Mosvatn Dam with 'Do you know auntie Kjersti?' Torstein Skinnarland would respond, 'No, but I know her brother.' With such homely introductions, the mission could proceed. These Skinnarland brothers, at enormous risk, would more than do their bit. By such seemingly small acts of brave defiance, tyrants are laid low.

Back in the UK, training piled upon training, followed by more training. The team was endlessly frustrated by the lack of action. Spring matured into summer, then slid towards autumn, no movement order came. The Commandos were a finely honed weapon but one that, unless tested, is apt to lose its edge. Wrong weather, wrong technicalities kept them in Scotland.

Meanwhile, they needed just the right kit. None of them had any illusions about the demands of just surviving on the Vidda. As it was their mission, they were provided with funds to shop around. This spending spree extended to ski equipment, marching boots, clothing, compasses and sleeping bags. Lillywhites of St. James (founded 1863), initially a supplier of cricket gear begun by a notable line of sportsmen, only had Alpine skis and boots, wholly unsuited for the cross country or Nordic style skiing the Commandos would be undertaking. (Nordic is a form of skiing in which the toe of the ski boot is fixed to the binding in a manner that allows the heel to rise off the ski, unlike alpine skiing, where the boot is attached to the ski from toe to heel. 'Telemark' skiing combines elements of both Alpine and Nordic styles, using the rear foot to keep balance while pushing on the front to create a carving turn on downhill skis with toe-only bindings. This would have been the natural, 'house' style for team *Grouse* members.)

The available stock of sleeping bags weren't much cop either: 'The sleeping bags had to be filled with down, and the stitching could not penetrate the area where the down was located. The inner part, that is, had to be divided into compartments, and each compartment had to be filled with down and stitched separately. At the same time, the inner layer of the bag had to be resistant to water, the outer layer capable of shedding water, and yet both layers had to breathe too.'[5]

Finally, they managed to find a bedding manufacturer located close to Trafalgar Square who, though they'd never stitched a sleeping bag, were keen to have a go – Poulsson's ample supply of ready cash certainly helped oil the wheels of patriotic commerce. Trial and error ensued and production at length proceeded. Happily, the Norwegian Government in Exile had established an ordnance depot at Dumfries. More familiar with the rigours of their own climate, artisans there were able to produce or make rucksacks and gloves, together with sheepskin-lined boots. Nordic ski equipment gear was finally sourced in Iceland (then under benign Allied occupation). Norwegian emigrés still fleeing across the North Atlantic brought in more useful items. *Grouse* was good to go, but there was as yet no 'go'.

For these eager members of *Grouse*, there were two false starts flared during September. Twice the team huddled, loaded with their kit at RAF Wick. On both occasions they shook hands with Tronstad and Wilson, only to be stood down. Each time, the Halifaxes did manage to take off, roaring away into the late summer's evening, those northerly nights already lengthening. They got as far as the Norwegian coastline, so tantalising near for the exiles. First time, an engine burnt out, forcing the pilot to abort and second go, dense fog over the UK compelled their return with a seriously touch and go landing in near impenetrable murk. Such tensions fray nerves and blur the sharpened edges of so many months of gruelling preparation.

Poulsson's frustration comes through very clearly in his diary entries:

> If you think we have left you are damned wrong. Today is Sunday, and we came here last Monday. A week's waiting for fine weather which never comes. Otherwise, it is alright here – the house is full of FANY's [First Aid Nursing Yeomanry].
>
> 12th September: There is a red light today and we hope for the best. We are now ready to start!
>
> 27th September: Of course, we came back. Motor trouble. We have now been here four weeks, but today, Sunday, we are going to try again.
>
> 28th September: Another unsuccessful attempt. Fog in the North Sea. Devil take the lot! But tails up.[6]

Helberg, resident team daredevil, was bored enough to turn his frustrations into verse: 'Have you seen them, the lads / They who rush through the air / The heroes from Piccadilly / ever more than willing / to offer life and limb / for their soldier friends. But now they have lost their splendour / all have stopped their dance / depression rules the boys / for they were the Commandos' drudges / but now they can only bluff / over deeds of a bygone time / they got 30,000 for the jobs / and that the lads couldn't stand.'[7] Not exactly Wilfred Owen but we can sense his emotions.

October 18 provided a 'bomber's moon' and the green light flashed for a third try. This time everything worked, and the Hardanger Plateau finally lay around 10,000 feet below. Their designated drop zone was in wet ground by Ugleflott. Easy enough to pinpoint on a map but very difficult indeed from such an altitude and in the dark.

They would jump using a static line, one at a time, followed by a dozen substantial containers, each packed with weapons, rations, comms and general kit. The lumbering Halifax now came down to a thousand feet. The terrain leapt to life now in full, startling relief, ridges, valleys, gullies and scree swam into greater focus. Poulsson and Helberg knew the area – at ground level in daylight. Nobody could be sure where they were, but the dispatcher was yelling 'action stations' and the beckoning void in the plane's belly beckoned them.

Absolute discipline for the jump was essential. No room for hesitation, every man had to heave himself out in rapid succession. The plane was covering 180 feet (59 metres) every second, so a ten-second delay between jumps could mean a spread of 600 yards (550 metres) on the ground. With four parachutists, their combined spread could be over a mile! They'd practised all this of course, settling over the hatch, legs dangling, then negotiating a clean exit and manoeuvring the cumbersome chute and passing through, plummeting to earth for 16 feet (five metres) until the line snapped the chute open.

Then it was that strange moment of weightlessness, a kind of intoxicating liberation, free of the incessant roar of aero engines, no more decisions to make, no more red lights to strain for and, at last, home lay beneath you. No more false starts and fractured hopes. Home, the fissured, broken waste that was the Vidda, one slight miscalculation, just one piece of bad luck and the reunion could suddenly turn fatally sour. Poulsson went first, then Haugland, Kjelstrup and finally Helberg. *Grouse* was finally coming back to roost in the jagged surface of the hyperborean Hardanger.

Helberg in his account picks up the intensity:

> *Jumping In* Finally, on 18 October, we had some luck. In the most beautiful moonlight, we circled over the Hardanger Plateau to find the target for our jump, the extensive Lokkje Marshes at Ugleflott. We believed that the navigator would have no trouble finding them. It is hard to describe my state of mind just before jumping. A few hours earlier, I had been soaking up sun on a lawn in England. The joy of returning home blended with the excitement of the jump itself. As the plane descended from 4,000 metres to 300 metres, the terrain no longer looked so fine and flat. 'Action stations!' The sharp command gave us something else to think about, and a few seconds later I was floating down to the plateau. We quickly learned that the pilots had not found Ugleflott after all.
>
> We ended up in rough terrain, in boulders and rocky mountain slopes. But all four of us plus our 12 containers with food, equipment, weapons and ammunition landed nicely. Even our radio transmitter seemed to have survived the rocky landing. After a year in England, we were not quite used to the sharp, dry cold.
>
> *Mission Revealed* We soon got something else to think about, once our team leader told us about our actual mission. To assist in destroying the plant at Vemork, we were to find a good landing site for two British gliders with 40 Royal Engineers and to guide the men from the landing site down to Vemork.[8]

Nonetheless, though all came down safely, the drop had not gone to plan:

> ... [They] had landed at Fjarefit in the Songedal. In the middle of Hardanger Vidda, instead of on the extreme east

> side. The RAF had dropped them 30 miles off their intended dropping zone. It was always curious to Norwegians how incapable the British and American pilots were at navigating over mountains and forests. They were looking for towns, navigable big rivers, railroad lines, big electrical conduit lines, etc. In the Norwegian mountains there is only wilderness.[9]

Haukelid, who was not a pilot, is being over censorious here. Flying a plane in wartime over such a vast and blank canvas is a most challenging task and such mistakes, whilst potentially catastrophic, were inevitable.

They were indeed a distance from Ugleflott. The slope they had landed on was a boulder- and scree-studded obstacle course, it was a near miracle that none of the party was seriously injured: 'Luckily, the crosswind was not too bad, so we all landed safely and so did our containers. Even our transmitter appeared to have survived. And here we were in Norway, cold and inhospitable, but marvellous all the same. After a year in Britain, we were of course not used to the sharp, dry cold on the plateau, let alone its snowy desolation and loneliness.'[10]

At last, they'd reached home, wrong location or not and Scotland, their adoptive home, now seemed a long way off. This brief disorientation was perfectly understandable, Scotland might look like Norway, but it wasn't the same, winter there more dank than sharp-fanged dryness. Poulsson lit his pipe. At this point his team still believed their brief as trainers for home forces. He disabused them of this and as he calmly puffed, he laconically outlined their real task. He let them know their objective was to facilitate the destruction of the heavy water facility at Vemork and that the actual sabotage would be accomplished by British Commandos to be flown in later. They were the pathfinders.

It didn't take long for the team to take this rather startling revelation in their stride. After all, this was still very much fighting back and that was why they'd trained so long and so hard, why they'd endured so many frustrations and why they'd parachuted into dark winter skies onto some of the toughest ground in Europe. Helberg soon came up with a suggested landing zone for this larger team, the Skoland Marshes, close enough to the Mosvatn Dam, out of sight of the Germans while offering a relatively easy march to Vemork. Poulsson had been thinking along the same lines. These men knew their native ground. He advised that the sabotage mission would be carried out during the next full moon, so they had just under a month to undertake a full recce. The job at first light was to gather in their gear. Meanwhile, they'd huddle to bed down, just praying no more snow fell to shroud their scattered containers.

Four pebbles thrown down onto a vast, wind scoured beach, a blend of remoteness and audacity that would probably not occur to the Germans, who shied away from the huge nothingness of the Vidda. Despite the immense hardships, this team were at home here and this was *their* country which a ruthless and rapacious enemy had stolen from them. They might not yet be able to take it all back but what they were about to attempt could make a far bigger difference than they knew. Failure was simply not to be contemplated.

Snow was now their ocean, a glittering sea of whiteness, jewelled by a clear bright winter sun, dazzling and deceiving. It took a full two backbreaking days to locate and drag scattered supplies to their ad hoc base camp. Poulsson dispatched each in lone missions to different point of the quadrant. The weighty, inert cylinders were infuriatingly difficult to locate and many a floundering dash ended in disappointment. Ground was difficult and broken, boulders tossed in random heaps of ancient moraine, a warren of uncharted gullies and ravines. As they had not located their skis, they had to tab through the clinging slush which seemed to stick like molasses to the soles of their boots, making a trial of each step – sod's law their ski gear lay in one of the last containers they retrieved.

What they needed next was a fix on their actual location and they worked this out easily enough from their maps. They were a good ten miles (16 kilometres) west of where they should be at a place called Fjarfelt, which lies east of the Songa River and some 65 miles (105 kilometres) distant from Skoland Marshes. Not a problem. All expert skiers, they'd soon eat up the miles, but they were now burdened with all their supplies and kit, rations, guns and ammo, fuel, camping equipment, that damned heavy transmitter with charger and spare battery, even more damnably heavy Eureka machine, total weight around 650 pounds (293 kilograms).

One resource they'd not been able to locate was the paraffin for their primus stoves. This was a serious blow, without it, no warmth. They couldn't just make a dash over high ground to reach their target zone but would have to wind a much slower way along the valley of the Songa where they could at least hope to find firewood. The longer their travelling time, the more rations they'd consume. These were tight at best and 30 days without any hope of resupply was tricky – they couldn't risk begging food from friendly locals, security was still their prime concern. But nobody had said this mission would be easy.

Each man would have to hump his maximum lift of 65 pounds (just under 30 kilograms). Tabbing cross country in such conditions, you couldn't expect to bear more. Given the total burden, this implied each man would have to go and come back three times, shifting the heavy kit by relay, even allowing for what they had cached at base camp for recovery later. It was now 22 October. Time, like everything else, was a luxury available only in limited supply.

Carrying these near crushing loads, they slogged through wet, cloying snow that coated the undersides of skis. Nordic skiing, the Telemark discipline, is a well-developed cross-country method which utilises a diagonal stride technique and these men were all very good at it, but such conditions, however fit the skier, are exhausting. Add in the weight and they're utterly debilitating. Had it been colder – and, this time of year it should be – then the going on frozen wetlands would have been easier, but winter wasn't yet wintry enough, so they slogged over broken shore lines and deep drifts of soaking slush Any false step, and there were many, meant they could be up to their thighs or waist in freezing water. Being wet wasn't just uncomfortable, it could be fatal.

It wouldn't have been so bad if the march had been just that, from A to B, even with the burden each man laboured under. But they had to go back and forth several times, over the same difficult ground, damp if not soaked all the time, which was even more dispiriting. Even when darkness descended after short winter days, they then had to hack down sufficient wood, every bit sodden, if they were to build a fire. This was like extreme camping enlivened by nerves and the constant fear of discovery. *Grouse* was the stiletto, not the broadsword. They had to remain not just invisible but unsuspected. They were ghosts, invisible to the foe. A moment's bad luck or involuntary carelessness could be the end of them. And, if they failed, whoever picked up the torch and came after would have a much, much harder job. One major reason they might succeed was that the enemy would not suspect that anyone would be daft enough to try.

Around the spluttering, guttering flames they huddled, trying to coax some warmth into their stiff limbs and dryness into wet gear. They could have been their own Viking ancestors, alone in an empty, hostile vastness. But no Norseman, however audacious, would have attempted their current enterprise, their forbears were motivated by profit rather than patriotism, so it wouldn't do to push the comparison too far. Yet it was that outward urge, that unquenchable spirit which still ran in their veins, the same physical toughness, stoical philosophy and unmatched courage. They'd be needing all of it.

It wasn't all desolation. The Hardanger was dotted with timber cabins and summer shielings (*saeter*). Whilst scarcely luxurious, strictly basic, these shacks, abandoned for the dark months, did offer shelter and now and then a forgotten stash of foodstuffs, generally no more than a couple of cans, but bounty to men on a very limited and meagre diet. Most of their waking thoughts focused on getting enough to eat.

They were very fit, about as fit as fit gets, but their gruelling marches demanded a high intake of calories and protein; they certainly were not getting enough. In one blessed hideout they came across a supply of frozen hardtack. Avidly they hacked the meat into slices with a chopping axe, then boiled it up in a heated plan filled with snow. The place also housed a decent supply of dry firewood. By a miraculous twist of fate, they also found a sled. This would greatly ease the burden of moving supplies, and nobody could accuse them of stealing – the toboggan actually belonged to Poulsson, a childhood treasure vanished during the war's early days!

In fact, this trail of huts was their salvation, mitigating the dreadful rigours of nights out in the open. They freely made use of whatever they found there and technically none of it was theft. Convention on the Plateau was that you could use other people's stored provisions in extremis, as long as you intended to replace these at a future date. Clearly, to all loyal Norwegians, serving your country counted as well.

Even titans can fail, and the intense strain imposed by mountainous loads and unyielding cold began to take its toll. Poulsson suffered a painful boil on his left hand, which he attempted to rest in a sling when not skiing. Such physical manifestations of this unhealthy cocktail of exhaustion and undernutrition inevitably multiplied. The weaker they got, the less distance they were able to cover, so more journeys were needed. Helberg recalled that on one particularly bad day, they only managed to cover three quarters of a mile.

By now their rations had dwindled to a below subsistence diet of quarter of a loaf of dried meat (pemmican), which was ground down into a paste, mixed in with fat and some dried fruit, compressed into portable cake, nutritious but not enough. Very small portions of oats, flour, hard biscuits, a meagre smear of butter, a knob of cheese and slice of chocolate, completed their daily allowance. Centuries before, the medieval English archer, 'men of the grey goose feather', needed a daily intake of 3,000 calories to maintain even basic fitness on campaign. Adult males require between 1,600 and 2,000 on civvy street. It would be difficult enough for most of us to function on team *Grouse's* intake, even with our sedentary lives.

Their sapping trek continued. Day by day they were inching closer to their objective, the Skoland Marshes. By 30 October, they'd been marching for a dozen agonising days but had at least reached a point seven miles (11 kilometres) south of these wetlands. They'd also come across yet another abandoned shieling at Reinar, no food but the compensation of a ready supply of dried firewood.

Roughly, they had only a fortnight before the anticipated para landing. So far, they'd not made radio contact, hadn't identified a decent landing site or recce'd the plant itself. And, of course, they needed to find food.

Food. Their limited rations were close to running out. The continuance of the mission demanded they get sustenance. Some supply still remained in the deserted shieling they'd occupied briefly on the march. Helberg (Dumas would have so loved him), volunteered to do a solo run back to garner whatever was there to be garnered. No mean task, a round trip of 50 miles (80 kilometres) through a morass of soggy, clinging slush. Poulsson and Kjelstrup would recce the proposed landing site while Haugland, now alone in their cabin, would sort comms. They'd return in three days. Nobody had it easy, Poulsson went through the ice trying to cross a half-frozen river, potentially life-threatening, and a second wet gap defeated their efforts completely. They had no choice but to return to base, soaked, frozen and very hungry.

Comms were also proving tricky. They'd not been able to establish contact with London on their march. This was vital. Clearly, if Tronstad and SOE heard nothing, they would assume the worst. Haughland set up his antennae at the hut but immediately ran into a serious hitch. He could establish contact, but the battery ran out each time just at this crucial juncture, dangerous as well as infinitely irritating. It also meant finding a new battery. If they couldn't get comms working, all their efforts thus far would count for nothing, a truly terrible prospect.

The good news was that Helberg returned laden with booty: flour, meat, oats and dried fish. They would at least eat. Poulsson recorded admiringly of his boyhood friend, 'He proved the old Norwegian saying, "a man who is a man goes on until he can go no further – and then goes twice as far."'[11] The Viking shades must have approved. But still no comms.

Fed adequately if not well, they trudged on a further short day's march to a different hut, standing on the shore of Lake Sand, just eastwards of the Skoland Marshes and a well-located operational base. They were close to the Mosvatn Dam. The keeper here Torstein, was Einar Skinnarland's big brother, active in the resistance as well as being an old pal of Helberg's. Torstein was staunch, he'd do whatever he could. He might just (hopefully) be the man to resolve their problem with batteries.

He was and did. On 9 November, team *Grouse* finally established contact with Baker Street: 'It was eerie in that solitary cabin in the wilderness to listen to a man in a city so far away. Knut Haughland lay in his sleeping bag, with only his sending hand exposed over the set. Our first message was to let Tronstad and Wilson know that we were alive and well. It was short and simple: "Happy landing in spite of boulders everywhere. Snowstorm and fog forced us to go down valleys. Four feet of snow made it impossible with heavy equipment to cross mountains."'[12]

It was game on, and they could now report each day's progress in short cryptic messages, ever mindful of possible German radio finding devices.[13] Remoteness wasn't a shield from prying ears. The Germans, too, could be very good indeed at what they did. Haugland also tested the cumbersome Eureka device to ensure it still functioned – much would depend on this during any airborne landing (see following chapter). They could now confirm to London that there was only a single company strength unit, mainly Austrians in the vicinity, though more German reinforcements were anticipated. None of the outlying farms possessed telephones, all calls went through a small central exchange, easy to deal with.

Their chosen landing ground was adjacent to a track skirting the marshes, around three miles (five kilometres) from the Mosvatn dam: 'Nice flat ground without trees or stones, approximately 700 yards [640 metres] in length'. Even better, the ground was completely hidden. Part of the original planning had the Commandos, once safely landed, cycling down to their target. But the snow was now too deep and the raiders would have to tab the distance, though they'd probably get away without snowshoes. Poulsson estimated the approach could be accomplished in around five hours.

Haughland was an expert radioman, but he'd been out of the game for a while, not to mention the fact his exposed fingers operating as though by instinct outside the cocoon of his sleeping bag were very nearly frozen stiff. Every operator has his own style or 'fingerprint' and his handler in London (safe and warm), would detect any variation. This could trigger alarms bells implying the agent had been captured and replaced by an Axis imposter, or even turned. Haughland's 'style' didn't seem to match what he'd left on record. A series of security checks, as frustrating to the man in the field as the fatuous questions most of us experience today when trying to speak to our own bank. Knut passed all these, but suspicions were still acute.

Q. 'What did you see walking down the Strand in the early hours of January 1st, 1941?
A. 'Three pink elephants.'

Right answer, panic over. This whole rigmarole, necessary as it was, must have struck Haugland, near frozen in his down-filled bag, isolated in the distant tundra of the Hardanger, as completely surreal.

Meanwhile, their daily struggle to survive continued. By now, all were suffering from undernutrition, which technically is an insufficiency of calories, proteins, essential vitamins and minerals. Symptoms, which from time to time they all exhibited, might include loss of fat and muscle mass, hollow cheeks and sunken eyes, swollen stomach, delayed wound healing, dry hair and skin, fatigue, difficulty concentrating, depression and anxiety. The job they'd taken on, its duration and intense physical demands, obviously exacerbated the problem. Men less fit, less disciplined, less motivated would have failed long before.

Poulsson was resident chef, a very arduous post:

> Preparing meals in that hut was difficult, since the room with the stove in it had no windows. It was dark, because we hardly ever had any lights and as a rule, we couldn't leave the door open, the way it was blowing and drifting. There was enough snow coming in anyway, through all the cracks … we had a long antenna mast made from planks which we nailed together, and it blew down time after time. We made fish soup – good soup too – out of dog food, some old, dried

> fish we found stored in the loft. Torstein Skinnarland also provided us with food. We were trying to eat our fill to regain our strength after the hard going on insufficient rations during the advance. But it was like a drop in the ocean![14]

An unexpected bonus followed when Haughland came across a lost sheep, penned in a ravine. The sheep's bad luck soon turned worse as it was quickly on the menu. Poulsson was able to deploy his latent chef's talent to spectacular effect, creating a mouth-watering sheep's head broth, compelling aromas wafting around the tiny hut, a feast for hungry fellows so long deprived. But then he dropped the kettle, scattering this eagerly anticipated fare across the timber floor. Everyone suffered a sense of humour by-pass and they were all soon scrabbling like hounds to grab what they could with their fingers as precious juices seeped through floorboards.

One thing they didn't need was visitors. Once they were encountered, however innocently, the chances of the enemy finding out increased radically. SOE's instructions from Baker Street were unequivocal. Anyone who stumbled across their hideout, unless it could be proved these were 'good' Norwegians, was to be permanently silenced. Poulson and his team were very aware of the risks, but they had very real qualms about murdering fellow Norwegians.

Two men turned up and this pair immediately set scalps tingling. One was apparently from Oslo, who just happened to have been alerted by the ad hoc antenna mast. His companion, a man called Brorusten, was known to both Poulsson and Haughland. It seemed that both were 'Good' Norwegians. This was difficult, the very idea of assassinating fellow citizens in cold blood was, rightly, chilling.

Ray Mears in his first-class account reinforces the point that the team weren't strangers here, their friends and families lived nearby, this was their home turf, yet they had to avoid all contact. To do otherwise was to invite the wrath of the SD or Gestapo to descend with all its frightfulness on their loved ones. That didn't bear thinking about and they really could not afford to add on additional risks. And not all Norwegians were by any means loyal to the Government in Exile, the odious Quisling still had far too many supporters, as zealous as their Nazi satraps. Brorusten and the man from Oslo went on their way.

Grouse had reported regularly on the strengths of the enemy garrison outposts in the vicinity. It was fairly obvious these were not front-line troops. Hitler's crusade against his former Bolshevik ally was sucking legions of men into the maelstrom and denuding resources left for garrison duty. Nonetheless, by now the German forces stationed in Norway had mushroomed to 300,000 and elite troops were rotated through Telemark but (happily), they didn't stay. Rather more alarmingly, Einar Skinnarland advised that German civilians had been identified in Rjukan. He was nigh on certain these were Gestapo – they didn't look much like scientists. Sappers were rolling out extra mines around the plant and more 'brass' were to be seen prowling the premises, no doubt with increased security very much on their minds.

With the German consul from Oslo in tow, the chief of the serious brass, Von Falkenhorst, toured the plant during October. He didn't need to be Napoleon to work out the risk British Commandos posed to his beloved jewels. Effective as all the recent strikes had been in their limited way, they clearly headlined the fact that Vemork must be high on any future hit list. The General wasn't aware of *Grouse*, if he had he'd have been a great deal more nervous.

These 'gangsters' were a force to be reckoned with, Von Falkenhorst harangued his audience. It's quite clear the general had a real admiration for their derring do and ruthless professionalism. He went so far as to demonstrate silent killing on some of the guards by way of instruction and listed the saboteurs' arsenal of tommy-guns, silenced pistols, fighting knives, chloroform, grenades, knuckle-dusters and the whole of Messrs Fairbairn & Sykes cabinet of horrors. He further cautioned his audience (by now probably bored stiff), that Commandos might very well adopt 'civvies' for the purpose of infiltration. Having bruited all these perils, the General confessed he didn't have adequate resources to install an adequate garrison and they'd have to make do with additional mines.

So, yet more mines were laid, scattered around the huge penstocks sweeping down from the heights above, searchlights were installed on the roof, an additional MG nest was set up, concealed in an otherwise innocuous looking hut by the entrance, and new booby traps added. Crucially, the cliff fronting the plant, descending so sheer into the deep dark valley below, was believed to be naturally impregnable, an oversight that was to have quite startling and decisive consequences.

Added security covered the single narrow bridge, gated at one end and regularly patrolled. We can be sure Von Falkenhorst was still uneasy. He knew routine garrison duty, so very far from any front line, was infinitely tedious. Complacency was inevitable. The place did look immensely impressive, carved out of a giant rock shelf with a vast chasm below. Besides, they were 150 miles (240 kilometres) inland, Commandos naturally preferred to strike at coastal targets Who would be mad enough to try?

In all fairness, at this stage it hadn't occurred to Tronstad or Wilson back at Baker Street to contemplate scaling that great slab of cliff, this would come later. Their planning focused on a conventional assault across the bridge. The plan was to kill the guards on the crossing silently, scale the remaining height up to the railway line, then cut through the mesh fence, dealing with any sentries. The saboteurs would be in a factory yard lying between the powerhouse and the plant itself. Enemy comms would have been knocked out before the assault team went in, they'd kick down the basement doors and plant their charges. Any Norwegian civilian workers would be led out of harm's way.

November 17 and the moon was out, casting its mellow light over the harsh profile of the Vidda, a bomber's or, in this case, Commando moon. Haughland transmitted: 'Lake covered with ice and partly covered with snow. Larger lakes are ice free. Last three nights sky absolutely clear with moonlight. Temperatures about 23 degrees Fahrenheit (-5 Celsius). Strong wind from north has died down tonight.'[15] Beautiful weather.

Act I of this drama, despite the characters many hardships, had gone well, very well all things considered. Tronstad and Wilson in London, the impresarios, would have high hopes of Act II – Operation *Freshman*, but such hopes would turn horribly sour and impose a far heavier burden on those four brave Norwegians, established on the Hardanger Plateau. They did not yet, mercifully, know this, but by far their greatest test lay ahead.[16]

Helberg leaves us a fast-paced account of their adventures:

> We also were to reconnoitre the plant in advance and obtain all information about its German guards. The landing site could not be too far from Vemork. The British would not have skis, which meant that the site would have to be close to a road. The Skodal Marshes, 5 kilometres west of the Mos Lake Dam, were a logical choice. They are on both sides of the road between the dam and Rauland. Some Tough Going The operation was to be carried out during the next moon period, which was in one month. Unfortunately, we had landed 15 kilometres farther west than planned, at Fjarefit in the mountain range east of the Songa River. It is roughly 100 kilometres from Fjarefit to the Skoland Marshes. We had to pull 300 kilos of the equipment we needed, including food, skis, our radio, and weapons and ammunitions. This would have worked out well on good snow.
>
> But in October we could not even expect snow for skiing. The snow that was on the ground melted the next day in the sun. However, we did not dare to go without taking our skis. If it should snow, we would be completely stuck without them. It took two days to collect our things, to pack what we would take, and to hide the rest in a depot. We had no time to spare. During the third night, heavy, wet snow began to fall. On 21 October we began marching to Mos Lake. For more than 14 days, we struggled in rough terrain and in indescribably bad snow. On good snow, an experienced skier would have covered that distance in a day or so. We just about exhausted ourselves to make it in 14 days. The Songa and Bitdal Rivers were not yet covered with ice, which meant that we had to struggle along steep birch-covered slopes.

> Each of us carried more than 30 kilos, and that was only half of the equipment, so we had to backtrack and walk each distance twice. Under such conditions, we were tempted to leave behind as much as possible, either in the depot or along the route. We soon realized that we would have to resort to the strictest rationing to make the food last for four weeks. This was our daily menu: 1/4 plate of pemmican, one handful of oats, some crackers, and some chocolate. Because of the toil, our need for calories was much higher than normal. Each day, we became weaker and weaker. Toward the last day we were not able to cover more than 2 kilometres.
>
> Fortunately, there were many cabins along the Songa and Bit Valleys, so we did not have to camp out. If we had had to sleep outdoors, I doubt we would have made it. On 10 November, we reached the Sandvass cabin, about 10 kilometres south of the Skodal Marshes, where we were to live until the glider party's arrival. On the trek from Fjarefit, our radio operator tried to contact England, but succeeded only once due to a defective battery. To get a new battery and to supplement our scanty food supply, I went down to the well-known skier, Torstein Skinnarland, who was the guard of the dam. Through him, we got all we needed, and from then on, the radio link worked smoothly. Skinnarland fed us information on Vemork, and after two of us had been down there to do a little spying, we were well prepared for the assignment.[17]

In reality, the 'assignment' was just getting underway.

At this point it's worth reflecting on the actual weapons the saboteurs might use. SOE might have given rise to James Bond, but specialized weaponry was the exception rather than the rule. It made for more operational sense for agents to be able to utilize Axis hardware which was likely to be easier to obtain in occupied territories. Obviously, there were exceptions such as the De Lisle Carbine and Welrod pistol, specifically designed for SOE, but these were relatively rare exceptions.

Basic and robust guns were to be preferred, and the crude, workaday Sten which has had its many detractors (this author included), remained a staunch favourite. Another stalwart was the US .45 Thompson sub-machine gun, the famous 'Tommy Gun' beloved of Chicago gangsters during Prohibition, a robust, well-designed gun, terrible at close quarters, though heavier and more cumbersome than the Sten. Most agents received training on standard Axis weapons, German MP 38 & 40, Italian 9mm Beretta, Walther & Luger 9mm pistols.

Given the overall weapons shortage in the UK 1940–1941, a variety of small arms were bought in from overseas, Colt 1911 .45 and .32 pistols, Spanish Llama 9mm, .38 S&W and Colt revolvers. More exotically, SOE bought up a consignment of 8,000 Ballester-Molina .45 pistols (an Argentinian home-manufactured version of Browning's design for the 1911 Colt auto), a deal brokered by the US.

Various models of sniper rifle were used, the most common being the adapted version of the No. IV Lee-Enfield service rifle. Selected models found during factory testing to be particularly accurate were fitted with a wooden cheek piece and mounts for a scope – the No. 32 3 5x telescopic sight. SOE, like some Home Guard Auxilier Units – 'Churchill's Secret Army' – relied on .22 sniper models fitted with suppressors. My personal favourite Auxilier survivor is a pump action .22 Winchester, beautiful heft and balance and very nearly inaudible. A marksman could hope to kill his man at 200 yards and make good his escape.

For the nastier end of their business, SOE used some suppressed weapons such as the .22 Ruger auto, but these weren't regular issue. As fighting knives many SOE operatives might have prewar hunting blades, they simply brought with them or prefer the legendary Fairbairn-Sykes or more specialized narrow bladed stiletto types.

5

FRESHMAN

'I'm starving.'
'We're expecting more company tomorrow.'
'Yeah, 50 British Commandos. From the Royal Engineers, they land tomorrow night.'

Heroes of Telemark

Thank Heaven! At last, the trumpets peal
Before our strength gives way.
For King or for the Commonweal –
No matter which they say,
The first dry rattle of new-drawn steel
Changes the world today!

Kipling, 'Edgehill Fight'

When German paratroopers, the Luftwaffe's elite *Fallschirmjäger* had so spectacularly landed on the roof of the great Belgian fortress complex at Eban-Emael (10/11 May 1940) and taken the allegedly impregnable redoubt in a brilliant coup de main, the age of the airborne warrior had arrived and with a suitably Wagnerian flourish. Here was a wakeup call indeed: Britain was then lagging very far behind. Operation *Freshman*, Combined Operations' first major glider-borne assault, was targeted at Norsk Hydro. With the inestimable benefit of hindsight, this was a very bad idea and would end in catastrophe.

As the terse German communiqué after the failure brusquely confirmed: 'On the night of 19/20th November [1942] two British bombers, each towing one glider, flew into Southern Norway. One bomber and both gliders were forced to land. The sabotage troops they were carrying were put to battle and wiped out to the last man.'[1]. In fact, this was very far from accurate, but in the end, the result was the same.

With *Grouse* already inserted, prevailing wisdom insisted that the actual raid on the plant depended on a much greater force and that a mere handful of saboteurs could not hope to get into the basement, place the quantity of explosives needed and escape without detection. Combined Ops, working with SOE, had calculated that the minimum number of attackers would have to be 15. Airborne Division agreed to commit to providing twice as many. So, hopefully both, but at least one, group would land safely, make it to the target and be in sufficient numbers to fight their way in. Realistically, their chances of getting out were minimal. This was, to all intents and purposes a suicide mission. At least that aspect of the overall planning proved horribly accurate.

The preferred November slot was rapidly closing and any list of alternatives just as rapidly dwindling. One option was to land a much larger force, ten times greater in numbers but delay the attack until December or January. Time was a luxury the Allies didn't possess. That costly success of *Musketoon* had prompted the occupiers to beef up security all round. Vemork was the jewel in their industrial crown. *Grouse* had reported near the end of October, confirming that an additional German infantry company had arrived in Rjukan with two more likely on their way. Furthermore, a single platoon was now based even closer to the plant in Vaer School at Vemork, less than a mile (1.6 kilometres) from the site.

More Germans, specialist Alpine troops, were billeted at Skarfoss Dam, 3 miles (5 kilometres) off and directly blocking the raiders' intended approach. At the facility itself, the penstocks were shrouded in barbed wire and electrical alarms, with discharging devices which, in the event of damage, would channel rushing water down over the cliff. Landmines may have been dug in and wires set up to frustrate aerial approaches. Von Falkenhorst, while he may not have achieved the garrison strength he'd hoped for, was taking the threat very seriously.

More radio messages from *Grouse* sent, as mentioned previously, on 19 November, cautioned that there was too much snow for an attacking force to move by bicycle as had been (rather optimistically) hoped. This meant that to land and mount an attack on the same night was now out of the question. It also clearly meant the twin Horsa gliders would have to be camouflaged during daylight hours to avoid detection from the air.

Despite these mounting difficulties and ever-lengthening odds, preparations continued as before. This job was perceived as being simply too vital to abandon, a view supported by Major-General Browning[2] in charge of airborne forces, together with the executive officer, Lieutenant-Colonel Mark Henniker, Commander Royal Engineers.[3] What to do about the gliders though? Colonel John Turner, the Air Ministry's concealment oracle, was consulted. His reply posed fresh problems. To conceal such a large piece of kit it would effectively need to be stripped down, wings taken off, together with fins, tailplane and landing gear. This giant Lego dismantling would have to be undertaken by inexperienced soldiers in the dark after landing and yes, the strains of said landing could distort the frame to make an already extremely difficult job much harder still.

This naturally demanded extensive practice. There was no time for such practice. Moreover, you could only hide such a huge beast where nature was prepared to lend a hand, ideally amongst small trees and bushes. On the Hardanger, nature didn't tend to lend a hand. Netting was needed to complete the camouflage, and this weighed in at about 600 lbs (272 kilograms) per glider. Besides, nobody had actually tried to hide one this big before. Having snow on the ground just made things that bit trickier.

Another problem with the air lift arose. Ageing Whitleys simply lacked the heft to tow two heavily laden Horsas through dark, frozen winter skies. Such a task demanded the mightier Halifaxes, and these were in short supply. Those who had them did not relish relinquishing their prizes. Now, if Bomber Command was to supply the planes should they not also provide their crews, already trained? On the other hand, if it was the Ministry of Aircraft Production that was to find aircraft, then 38 Wing would seek to train up their own crews with the aid of ground staff temporarily detached from Bomber Command. As it happened, there was a trio of Halifaxes temporarily idle at Netheravon 'drome on Salisbury Plain. One aircraft lacked numerous items of kit and essential parts, another was more or less good to go, though it presently lacked armament and the third was fully operational. None was fitted with the vital Rebecca gear; this required a commitment of no fewer than 200 man hours to install.

Make do and mend remained order of the day and training at Netheravon got underway for both tugs and their glider-borne brethren. Try as everybody might, it was impossible to duplicate actual conditions as found on the hostile terrain of the Hardanger plateau. Notes of urgency multiplied as fresh reports of tightening security around the Vemork plant rolled in. These included the possibility that a number of German 'civilians', recently drafted in, were more likely to be Gestapo agents. Senior Axis officers toured the plant itself. The Uranium Club's crown jewels were thoroughly protected. Von Falkenhorst was clearly acutely aware of his responsibilities and as noted earlier, while giving his forces a rousing speech had praised the effectiveness of British Commandos and stressed the need for constant vigilance.

Initially, the plan was that the raiders should launch their attack across the suspension bridge, clamber up to the railway, head west to breach the security fence then disperse to deal with their specific targets. Exactly how the attack would be carried out would be determined by close reconnaissance on the day in liaison with Team *Grouse*. Those already on the ground would mark out the landing ground some 12 miles (19 kilometres) from the factory.

Depending inevitably on prevailing weather conditions, their march to Vemork should be accomplished within five or six hours.

Those Commandos now completing their intensive training had undergone final selection. Their cover was an ingenious one. They were told this was an exercise readying them for participation in 'The Washington Competition', which involved pitting them against American personnel in several strenuous challenges. Only later were they told their real purpose was hazardous, behind-the-lines sabotage with a high risk of death or injury. As ever, those who declined could return to their units without censure or shame.

Heavy Water was codenamed 'Lurgan' though none of the volunteers yet knew anything of the importance of their mission. Training was intensive and tough. Volunteers were divvied up into two separate groups, each comprising one officer and fourteen troopers. They were provided with specialised kit, clearly intended for use in cold weather conditions with heavy duty packs designed to cart up to 80 lbs (36 kilograms) of gear. Next, they were ferried to Snowdonia for mountain warfare training, followed by meticulous instruction in the use of explosives.

Some must have guessed the nature of their intended target when they were moved up to Fort William to practise attacking a hydro-electric plant. Another clue appeared when they entered the Sunlight Soap factory at Port Sunlight, an installation which offered a range of condensers very similar to those they'd be blowing up for real at Vemork. More and even more intense training followed. At a Commando school they were able to get the feel of their intended target, working in a full-scale mock-up of the place with replicas of the plant to be destroyed. They became familiar with the location and surrounding terrain without ever being told where they were headed. They worked up their escape plan and, should they encounter opposition, they were thoroughly educated in the lethal arts preached so effectively by Fairbairn and Sykes, becoming as fast and accurate on the draw as any western gunfighter.

Only in the final stages, surrounded by a wall of secrecy, did they learn of their actual destination, of the intention to infiltrate them by glider and, less prosaically, how to walk in clumsy snowshoes.

There would be a lot of snow. Once they had completed their mission, they'd split into pairs for the long, hazardous trek towards neutral Sweden and safety. Each of these pairs had a different designated route and their getaway plans came complete with silk maps, button compasses and Norwegian krone, all sewn into their battledress. They were left in no doubts as to the vital nature of their task and nobody was giving out any odds for their survival. A great deal of thought went into their projected escape route, none of which could disguise the tremendous amount of hardship, the significant nature of natural barriers and the fact every German in Norway would be hot on their trail.

Speed clearly was the key. Once they had penetrated the complex, two sections would blitz the power house and another pair focus on the electrolysis plant. A brace of glider pilots[4] would trash the offices whilst a second duo remained as part of an HQ group which would corral any workers out of harm's way. Estimated time from start to finish was 20 minutes. They'd be in standard '37 pattern battledress (white snow camouflage kit would only be worn if necessary) but without insignia, steel helmet, personal weapons and gear. Anyone who got hit was to be pumped full of morphine and left. None of these men was aware of the *Kommandobefehl* at this point. It was assumed that being regular troops in uniform, they'd be treated as PoWs. This civilised delusion would engender dire consequences.

Quaintly, their SOE shepherds would begin 'This is Piccadilly,' and the chorus would be returned, 'I wanted Leicester Square.' If a glider went badly off course but landed safely and without too many casualties, the officer in charge would have to try and ascertain their precise location, camouflage the glider, and then make a tactical assessment as to whether it remained possible to reach the objective. Otherwise, they'd head for Sweden.

This was all pretty ad hoc and subjective. Nobody was under any illusions as to the dangerous and unpredictable nature of this operation. The most comprehensive training, best planning and superior kit couldn't save them from the vagaries of weather or disguise the fact this was something the British had never tried before. There's an old military saying that no plan ever survives contact with the enemy. In this case the initial enemy was November darkness over Norway and the unforgiving harshness of the Hardanger Plateau.

Combined Ops and SOE, working hand in hand (though Wilson always had significant doubts) saw their job as crucial. The war cabinet agreed, but persuading Bomber Command was a whole different matter. Conventional forces, army and RAF, had a rather dim view of these Special Forces johnnies, mavericks and cowboys whose function, as they perceived it, seem ed to be to distract and detract from the main war effort. However frustrating for the Baker Street Irregulars, they were Oliver Twist to RAF's beadle.

Nonetheless, there was some good news on the flying front. Two Halifaxes with three gliders had arrived at RAF Skitten from Netheravon on 16 November, but both aircraft needed work before being signed off as fit for purpose. Group Captain Tom B. Cooper, having delivered his first two planes, flew back to Salisbury Plain to taxi the third. Thus, the operation couldn't be launched straightaway and weather conditions were fickle. It might, however, just be possible to carry out a dummy run.

More good news was the relatively upbeat assessments being radioed back from *Grouse*. Weather conditions were favourable over the designated landing zone; whilst snow made cycling impossible, an approach by foot was still viable and the team supplied fully accurate assessments of enemy strengths and security arrangements. 'Beautiful weather' was the cheery forecast, night skies were light and bigger lakes like Mostvatn were still free of ice. It all seemed like a 'go'. Norwegian metrological officer Lieutenant-Colonel Peterssen was at Wick to assess the forecasts. Conditions were clear enough for a recce flight on the 17th, but the planes weren't ready. Next evening, they were, and the reconnaissance flight passed without incident. So far, so good, but Peterssen still wasn't happy about the prospects for the night of 19 November.

This was where it all started to get shaky. *Grouse* advised that conditions over the landing zone were good, but Peterssen was bothered about wind speeds. It's one thing just to fly in a hefty four-engined aircraft, but very different when the plane is towing such a cumbersome load. The meteorologist stuck to his guns during an afternoon review. He was unhappy with the available data and felt better conditions would be likely to obtain in a few days' time. But *Alea iacta est*, as Caesar is supposed to have observed, this mission had assumed its own momentum, men were keyed up, tooled up and ready to go. And go they would; but Peterssen's pessimism would be tragically vindicated.

Meanwhile there was mini drama on the ground. 2nd Lieutenant Mike Green was injured in a negligent discharge incident at the 11th hour. This was a grievous blow to the young officer, after all the effort, training and commitment. Moreover, his accidental survival would haunt him, though he did have a distinguished war. Lieutenant David Methven replaced him for the op, such are the vagaries of war. At 17.45 hours that evening, the first of the Halifax tugs roared into the air, followed 15 minutes later by the second. *Freshman* was on. (For the full order of battle see appendix 2).

Tension and tobacco fug clogged the air at Wick. Henniker himself was there and shouldering responsibility for the lives of all those men he'd just sent out to war. They were his men, he had selected and trained them and now he knew them, knew of their lives and families and hopes. Command brings a heavy burden, deepened by agonies of waiting. And it was to be a very long wait. It was around 23.40 hours when the first faint radio transmission was received. One of the bombers was seeking a bearing. Wick was unable to respond. Fifteen minutes later, a second signal came through with the enigmatic but highly disturbing message that their glider had been released over the sea. This was immediately followed by a confirming signal, giving a location.

Acting on this information, an air/sea rescue operation was scrambled. No ships were reported in the vicinity, which was 7 miles (11 kilometres) off the Norwegian coast. Weather was closing in, which could only further hamper all efforts. GCHQ and SOE were both involved but the situation only became more confused. Around 03.00 Squadron Leader Wilkinson's 'A' Apple touched down at Wick. Bit by bit the jigsaw pieces began to fall into place and the ugly picture began to appear. The tug had failed to locate the landing ground and so had turned around to abort and return to base, glider still in tow.

At 12,000 feet (3,658 metres) Wilkinson's Halifax had encountered stringy cloud but couldn't gain further altitude due to icing up. At this point the tow was around three thousand feet (914 metres) below, and presumably due to the weight of icing, it broke, the hapless glider spinning away to crash at some point over land. Cooper's subsequent reports to GCHQ suggest that navigation had been difficult from the outset and the Rebecca gear had failed.

Of 'B' Baker there was no news. Next morning *Grouse* reported that their Eureka sets had been functioning and whilst they'd distinctly heard the Halifaxes, they had seen nothing. A further search mission with a clutch of planes and one destroyer was sent out but nothing was found. Silence reigned. To all intents and purposes, that fog would stay swirling until 1945, when post-war investigation and trials finally revealed the truth, in all its ghastly detail.

Something indeed, a series of somethings, had gone terribly wrong. A mere 24 days before the mission took off, on 26 October, the *Kommandobefehl* had been sent out in Norway. The bad news would only get worse, and the full awful picture wasn't really put together until the end of the war and the hunt was on for Nazi murderers. 'A' for Apple piloted by Squadron leader A.B. Wilkinson (with Group Captain T.B. Cooper) on board, took off on 19 November towing its glider over southwest Norway, the Horsa flown by Staff-Sergeant M.F. Strathdee and Sergent P. Doig, carrying one Royal Engineers officer, Lieutenant D. A. Methven, with 16 ORs. Barely had they crossed the Norwegian coast than the Rebecca gear commenced playing up – this was in fact viewed as a major factor contributing to the unfolding disaster.

Nor could their landing zone be found, either on a first or second pass. Fuel was now running low, dangerously so. As the Halifax turned for home, dragging its heavy burden, the glider couldn't seem to rise above the cloud layer; this was around 40 miles (64 kilometres) northwest of the target zone. Ice began forming on the wings and tow rope, dragging both aircraft lower. Then, the vital umbilical suddenly gave way and snapped. Initially, Wilkinson thought the glider had broken free over the sea but after his plane touched down at 01.51 on 20 November, a reconciliation of courses and timings proved this misfortune had, in reality, occurred over land.

And 'B' for Baker? This was flown by Flight Lieutenant A.R. Parkinson of the Royal Canadian Air Force and his 2 I/C Pilot Officer G.W. de Geney (RAF(VR)). The glider crew comprised Pilot Officer N.A. Davies and co-pilot was Pilot Officer H.J. Fraser, both Royal Australian Air Force, Officer Commanding Lieutenant A.C. Allen RE together with fourteen ORs. From take-off there was radio silence until 23.41, when a message from the aircraft was received requesting a return course. Whilst the fix for this transmission could be calculated, nothing more was heard.

An extensive search was mounted in daylight by two squadrons from Coastal Command but yielded no results. It was only on the 21st that the British heard of the German announcement claiming all the Commandos had been killed in a firefight. Whilst this confirmed that both gliders had come down in Norway, the Axis assertion that there were no survivors rang hollow. On that basis, as Henniker posited in his report (the only real source available, see below), there appeared to be grounds for hoping some might have escaped. British operational orders issued on 15 November had been slightly ambiguous – when carrying and using weapons, the men must remain in uniform. If they ditched insignia and opted for civvies to facilitate escape, they had to dump their arms as well. Naturally, if they were captured in civilian clothing but still with weapons, they could be shot as spies. This smacked of wishful thinking. Assuming a safe landing and then a march to the target, they'd been instructed to retain their battledress but stripped of divisional markings.

What these blandly objective musings cannot conjure up is the terror aboard both gliders as they spun or plummeted to earth. The utter helplessness and bowel-churning fear that must have gripped every man. Out of control, no means of propulsion, on bad ground in a hostile land in freezing conditions, kit flying like oversize shrapnel. These were all very brave young men, but even the most sublime courage has its limits. Those who did survive the banshee wailing, tearing tsunami of impact as the gliders were ripped open like cereal packets spitting out their insides might, in the stillness which ensued, have thought themselves lucky. They would have been quite wrong.

At around 2.15 on the morning of 20 November the Hovland family who farmed land bearing their name in Helleland,[5] a small community just northeast of Egersund strung along the banks of the River Hedlandsana, were woken from their autumnal slumber. Trond Hovland, son of the local *lensman* or sheriff and an ex-copper himself, was in for a shock: the last thing he expected to see were two British Tommies in uniform who, fetching up at his door, managed to communicate that their plane had crashed a couple of miles away.

Trond and his father swung into action, obviously a rescue effort needed to be mounted but the nearest doctor was in Egersund and while there was a functioning phone link, the call must pass through a central exchange, which meant the situation couldn't be kept secret from the Germans. Of course, these decent Norwegians knew nothing of Hitler's diktat and assumed survivors would be treated as PoWs. Hovland duly informed the commandant at Slettebo camp near Egersund, advising of the facts and seeking help. By 05.30 a patrol arrived at the farm. One of the raiders was kept there under guard whilst the other was conscripted as a guide. A group of younger Norwegians, headed up by Arne Lima and being used by the occupiers as forced labour, were roped into the effort.[6] It was they who reached the crash site first, easily outpacing the lumbering Germans.

Given the impact of the crash landing, it was a miracle that any of the occupants had survived. Three had not and another trio were in a bad way. Arne and his companions had advised survivors to make a run for it, but they were a very long way from Sweden and with severely injured comrades to think about. Accepting capture and incarceration as PoWs seemed the only rational choice. They were at least able to burn all documents and other evidence that might be of use to the Germans who, once they'd breathlessly caught up, took over. Trond Hovland was ordered back to the lower valley to rake up more men with stretchers for the casualties.

By now it was 07.00 and Hovland received a phone call informing him a second plane had crashed. His father plus a Norwegian Quisling interpreter set out to investigate this new catastrophe while Trond, with a score of helpers, slogged back uphill to the first wrecked glider. Dawn exposed diabolical carnage. The flimsy plywood craft had splintered like a balsa toy, it seemed impossible that anyone could have survived. Getting the living, wounded and dead back down the treacherous path to the farm was backbreaking work but accomplished by mid-morning.

Here those living were tended by the German doctor who had by now arrived, laid alongside their dead mates in the lensman's office. Before long, they were transferred to an ambulance and the other survivors were packed into trucks. They seemed cheerful enough, but Trond Hovland never saw any of them again. Only later after the bitter truth leaked out did he and the others wonder if they could have done more. Could they have facilitated the escape of at least of some of the British, could more of them have been hidden at other farms? At the time, like the raiders themselves, they had assumed the Germans would act in accordance with the Geneva Convention. That they would in fact behave as barbarously as they did was a serious wake up call for many, if not all Norwegians.

Crash site no. 2 was the remains of the second bomber, piloted by Flight Lieutenant Parkinson. Wick still knew nothing of what had happened to this plane. A farmer living at his steading called Haestad had phoned Hovland as the nearest law enforcement officer and the lensman got into his car and motored to the farm. Here he learned another local man, Tonnes Drange, had witnessed the Halifax coming down sometime around midnight. From the sound of the engines the plane appeared to have been circling before it smashed into a bare hillside. Others working on the Oslo to Stavanger railway line had picked up the sounds as well. While the location was relatively close to a road, it was still on high, almost inaccessible ground.

As if the sheriff had not seen enough horrors for one morning, this was a scene from the apocalypse. The doomed aircraft had gouged a deep furrow, hundreds of metres in length before breaking up. Debris, wreckage and body parts were scattered in grim profusion. One crew member's corpse, an aviator, was relatively intact, he'd been thrown out on impact, the rest were savagely mangled. Nothing in the experience of these Norwegian farmers could have prepared them for such sights. There was no need for medical attention, just the ghastly chore of collecting human remains.

Germans arrived and simply flung these body parts into a swamp, so carelessly severed limbs were left standing like macabre sentinels. The Norwegians' horrified outrage was brushed aside until they pointed out the remains could pollute the water supply to Slettebo camp. That did strike a chord, so the mangled bodies were re-gathered and buried with no more dignity in a mass grave scratched from the mountainside. At least the locals were able to recover some identifying papers and kit from the site.

From the outset, Henniker had been conscious of the risk that one of the gliders might not make it to the target, but he had not imagined that both would fail. So far, one crashed glider and Halifax had been found. There might have been the hope that the second glider could somehow have got through. It hadn't. The Horsa had come down in Fylgjesdalen – a remote steading, its own private commune since 1887, located in a high, towering landscape 500 metres above the deep fjord below. In terms of its history, not much had really ever happened there. That had just changed.

A pair of small farms, little more than summer shielings, clung to lower ground on the north flank of the fjord – no roads connected them but from each, narrow tracks led upwards for a kilometre or so to a lonely mountain tarn and the steading at Fylgjesdalen. The farmer, his extended family and one elderly servant lived there, bearing the same name as their ancient land. In the wee small hours of 20 November, they were roused by three uniformed strangers who appeared out of nowhere. These men were all soaked, dishevelled and distinctly shop-worn. Whatever these apparitions represented it wasn't good news, like harbingers of death from old Norse legend.

Thorvald Fylgjesdalen couldn't understand the strangers. For all he knew they might have been German, but they were clearly in need of medical assistance, so he led them down to the Haaheller's farm below. Jonas Haaheller couldn't understand them either but had divined they were more likely to be British. All he could do was to put a call through to the lensman in Hole.

Thorvald Espedal was the sheriff but had been shunted sideways by the Germans who suspected he was too true a Norwegian to be trustworthy. Leif his son had been passed over for the same reason, so, rather bizarrely, the baton had been passed to Espedal's daughter, Leif's sister Hjordis. By 01.30 on 20 November, it was she who heard from Jonas Haaheller. Having briefed her brother, both agreed there was no way of bypassing the Germans who had a firm grip on Rogaland Police HQ. A doctor was summoned, and the assistant chief constable said he would come to Hole personally and the investigation/rescue party would cross the fjord together in a motorboat. Nonetheless, it was Espedal senior and the doctor who were first on the scene at Fylgjesdalen. By now it was late afternoon.

At last able to communicate with an English speaker, these three soldiers explained their plight, that their glider from Britain had broken free of the tow around midnight and had crashed with seven men killed on impact. Those who'd gone for help were the most intact survivors, the rest were still on the mountain. The Tommies had no clue as to their present location but wondered if they might yet escape to Sweden. They were a very long way from the border. Might they not then escape by sea seeing as they had fetched up near the west coast? These sympathetic locals pondered this but concluded it wouldn't be viable and that, if they helped, the heavy hand of reprisal would fall firmly on their tiny communities. Again, nobody present was aware of Hitler's order.

As short autumn's dusk drew in, Hjordis Espedal, Leirvaag, the ACC with Inspector Ravn Tollefsen reached the farm. Hjordis and Leif, together with the inspector, would climb up to the crash site whilst the ACC and doctor agreed to muster a rescue party. Local

Norwegians weren't shy about coming forward and several groups coalesced to begin the operation. It was bitterly cold; it took the two younger Espedals three hours to make the climb up to Thorvald Fylgjesdalen's alpine farm. More stiff climbing awaited them as icy gusts of sleet and snow closed in. In normal circumstances, nobody in their right mind would willingly undertake such a journey in the ice-bound blackness.

At this point a further trio of survivors, all in a noticeably bad way, had made it down to Thorvald's steading. Private White, despite losing an eye and several teeth, was able to act as guide, provide some details of the mission and describe those final terrifying moments in the whirling sky before the glider plummeted onto the rocks. The Horsa had come down nose first, crushing the cockpit and shearing off the flimsy wings. Kit and gear spilled everywhere, the whole terribly scene etched eerily in the light of a clear moon. The rescuers counted a total now of eight dead. Their deaths had not all been instantaneous. One had crawled some distance from the wreck and tried to apply a splint to his busted leg, shock and the cold had taken him. Another lay with his intestines ruptured, spilled guts frozen.

Three more survivors yet lived, each one seriously injured. There wasn't an awful lot these rescuers could do for them, moving them as gently as possible into sleeping bags and giving them hot drinks. More rescuers struggled up by 03.00, with yet more on their way. Those badly hurt were strapped onto stretchers and carted back down to the farm. The inspector with Leif gathered up all papers and documents they could find and burnt them. Shortly after, an explosion roared through the wreckage and set fire to the plywood shell. As the hull burned, the Norwegians could only drag the dead clear.

Getting the wounded, their sufferings eased by morphine, down to the fjord was a Sisyphean task. So far, the occupiers remained ignorant, the sheer remoteness of the place delayed their reaction. But it would come. Leirvaag, the ACC, received a phone call at the Haaheller farm from the Gestapo, who advised him that the security of these Allied PoWs was his responsibility and, should any abscond, he was in line for a firing squad. One thing about the *Geheime Staatspolitzei,* you knew they weren't ever bluffing.

When the Germans did arrive hey were mob-handed, two boatfuls of them, a mix of ordinary *Wehrmacht* with a senior officer present and security forces, led by Gestapo Commissioner Kuhn. Clearly rattled, the Nazis threw an armed guard around the tiny farm at Haaheller and arrested their prisoners. Sledgehammer and walnut, the wounded Tommies were in no condition either to resist or escape, all in a pitiable condition. They received no pity; put up against a wall, regardless of their injuries, they were then roughly loaded onto one of the boats, bayonets at their backs. The other victims brought down from the hill a distance away were chucked onboard in the same manner. The Norwegian police asked for the men at least to be taken below decks. This was refused, the boat motored back to Stavanger and no Norwegian ever saw these men alive again.

There now only remained the three still up at Fylgjesdalen and more SS troops under another Gestapo officer named Hartmann set off up the steep ascent. At the farm these casualties were clearly grateful to Hjordis Espedal for her decency and care. She'd remember one of them had a mouth organ, striking up a tune or two to ease their fear and suffering. But as soon as the Nazis swarmed in, she and Ravn Tollefsen were brusquely ordered away. Both were exhausted, but Hartmann still had work for them. They were to liaise with the arrival of another vessel from Stavanger, which would take off the remaining prisoners. Once across the fjord at Forli, locals were allowed to supply the captives with refreshment, but on Monday morning they, too, were moved to Stavanger and disappeared from view.

Hjordis Espedal and her neighbours had one last task to perform. On Tuesday evening, she was instructed to put together a work party, return to the crash site and bury the dead. These were to be interred high up, near to the scene. She protested that the bodies should be brought down and afforded a proper interment. Request denied. Nonetheless, the locals made sure to mark the site with a simple fence so that, after the war, the lost might be recovered and given a decent re-burial.

Kommandobefehl – I've mentioned this already but it's worth looking at the order in more detail here as the repercussions for *Freshman* survivors would be fatal and those responsible would be brought to account at war's end. This diktat was controversial from the start and the wording prepared after significant consultation with Third Reich lawyers. The response is said to have been prompted by several previous Commando raids against occupied Europe. Operation *Basalt*, carried out during the night of 03–04 October 1942, a raid on the island of Sark, resulted in several German casualties and it was said these men had been deliberately executed by the raiders, drawn from the Small Scale Raiding Force and No. 12 Commando, (which is disputed). The order itself stated:

1. For a long time now, our opponents have been employing in their conduct of the war, methods which contravene the International Convention of Geneva. The members of the so-called Commandos behave in a particularly brutal and underhanded manner; and it has been established that those units recruit criminals not only from their own country, but even former convicts set free in enemy territories. From captured orders it emerges that they are instructed not only to tie up prisoners, but also to kill out-of-hand unarmed captives who they think might prove an encumbrance to them or hinder them in successfully carrying out their aims. Orders have indeed been found in which the killing of prisoners has positively been demanded of them.
2. In this connection it has already been notified in an Appendix to Army Orders of 07.10.1942. that in future, Germany will adopt the same methods against these Sabotage units of the British and their Allies, i.e., that, whenever they appear, they shall be ruthlessly destroyed by the German troops.
3. I order, therefore: — From now on all men operating against German troops in so-called Commando raids in Europe or in Africa, are to be annihilated to the last man. This is to be carried out whether they be soldiers in uniform, or saboteurs, with or without arms; and whether fighting or seeking to escape; and it is equally

immaterial whether they come into action from Ships and Aircraft, or whether they land by parachute. Even if these individuals on discovery make obvious their intention of giving themselves up as prisoners, no pardon is on any account to be given. On this matter a report is to be made on each case to Headquarters for the information of Higher Command.

4. Should individual members of these Commandos, such as agents, saboteurs etc., fall into the hands of the Armed Forces through any means – as, for example, through the Police in one of the Occupied Territories – they are to be instantly handed over to the SD.

   To hold them in military custody – for example in PoW camps, etc. – even if only as a temporary measure – is strictly forbidden.

   This order does not apply to the treatment of those enemy soldiers who are taken prisoner or give themselves up in open battle, in the course of normal operations, large-scale attacks, or in major assault landings or airborne operations. Neither does it apply to those who fall into our hands after a sea fight, nor to those enemy soldiers who, after air battle, seek to save their lives by parachute.

5. I will hold all Commanders and Officers responsible under Military Law for any omission to carry out this order, whether by failure in their duty to instruct their units accordingly, or if they themselves act contrary to it.

Those survivors of Operation *Musketoon* were murdered because of this order, as were those involved in Operation *Freshman*. Norwegians participating in *Grouse/Swallow/Gunnerside* were therefore under no illusions. Yet the question remains, why did Hitler put out the order in the first place? October 1942 was three years into the war, so the timing seems odd. Various reasons have been put forward and no definitive answer has emerged. Historian John Warwicker amongst others has offered a very plausible explanation when writing about Home Guard Auxiliers, 'Churchill's Secret Army'.

Warwicker asserts that Auxilier Intelligence Officer ("IO") Captain Nigel Oxenden could have been the planner of a series of pinprick raids against German garrisons holding the Channel Islands – the only region of Britain they ever did get to occupy. The captain had been a pre-war resident. Whilst 1942 had seen major Commando activity at St. Nazaire and the unfortunate attack on Dieppe, the *Jubilee* disaster, smaller strikes remained the preserve of the Small Scale Raiding Force (SSRF), otherwise branded as No. 62 Commando. On 3 October a dozen Commandos (who may have included some Auxiliers), landed on Sark; Operation *Basalt*. This was classic small-scale raiding that we British have always loved, in the early nineteenth century this could have featured Hornblower or Lucky Jack Aubrey.

They were led by Major Appleyard. (John Geoffrey Appleyard, DSO, MC and Bar, was a Yorkshireman who began his war with the Royal Army Service Corps, RASC, and served in France prior to evacuation from Dunkirk. He then joined the Commandos leading SSRF till its disbandment in early 1943. That spring he joined Bill Stirling's 2nd SAS Regiment and died in July in a plane crash whilst actively involved in the Allied descent on Sicily.) The Commandos were taken in aboard MTB no. 344 and scaled the dunes to biff a small coastal hotel which housed a small enemy contingent. One German sentry was dealt with, thanks to Messrs Fairbairn & Sykes, while another five were captured, bound and led back to the beach as PoWs. The prisoners apparently didn't go quietly but raised a ruckus to attract attention from their comrades while struggling to get free of their bonds. It is said that Appleyard yelled that somebody needed to shut the captives up. Another officer added the postscript that escapees should be shot.

And they were, only one prisoner survived, the other four were left dead in the beach, allegedly with their hands still tied. The suggestion is that Hitler viewed this incident, minor as it may seem, as a clear indication that British Commandos engaged in illegal tactics and were not thus to be considered honourable opponents. Now, it is almost wholly conjectural whether this incident happened as described, or that it prompted the Fuhrer to issue his infamous order. But it does fit.

Naturally, the locals and activists from Milorg wondered what had become of the captives. Meanwhile, the Germans put out their short communiqué advising the Commandos had been wiped out in a firefight, which they obviously knew to be untrue. The resistance tried to dig deeper, and it wasn't long before rumours began to circulate that PoWs had been murdered, but no concrete evidence as yet emerged. Mark Henniker, who'd had his reservations from the start and who had endured that awful and fruitless vigil waiting for news that never came, was very keen to discover the truth. He didn't believe the German communiqué, it was too pat. He divined the Germans were trying to draw the British out into giving away more details. He felt the men should simply be posted as 'missing in action'. All details of the failed mission were to remain secret – a cruel twist of the knife for relatives.

Speculation and rumour abounded, and it wasn't only Combined Ops and SOE who were perturbed. The callous murder of all survivors also angered General Wilhelm Rediess and Terboven. They were irritated that these men had been dealt with summarily, not on humanitarian grounds but for the missed opportunity to harvest intelligence. It was a supreme and bitter irony that both Gestapo and SS were irked that the Fuhrer's order had been actioned so rapidly.

Rediess complained to Berlin that '...the Wehrmacht executed the survivors, so clarification is no longer possible.'[7] The Germans' haste to obey their leader had denied them the chance to accurately ascertain what the intended target was and whether, crucially, there were already forces in-country. Their savagery in fact aided the *Grouse* team. Von Falkenhorst gave clear orders that captured saboteurs were in future to be handed over to the security services for thorough investigation. Once milked of anything information of value, then they could be murdered at leisure.

By late 1945, the war was over but the search for what had happened to those survivors from both crashed gliders was just approaching its grim denouement. In mid-December, several

German defendants appeared in front of a military tribunal sitting at the Law Courts in Oslo.[8] These defendants charged with the murders of four as yet unidentified British soldiers were *Stabsarzt* Werner Fritz Seeling, *Hauptscharfeuhrer* Eric Hoffman and *Unterscharfeuhrer* Fritz Feuerlein. All pleaded not guilty to the charges. Major Burnett acted as prosecutor and the German defendants all had the benefit of legal counsel (who themselves had assistance with translation). This trial related only to the fate of those men captured in the wake of 'B' Baker's crash.

A key question for the accused was whether the captured soldiers had been wearing uniform or civvies. Evidence strongly suggested the former and that civilian replacements had been provided by the Germans when the captives were in transit between Stavanger and Oslo. Five PoWs were in fact sent to the capital while the remainder were to be dealt with according to the Hitler Order. Clearly, the accused were aware of the Fuhrer's order and felt bound by its directions.

One witness had shared quarters with Hoffman and related how the officer had described the deaths of these five prisoners: all had been given supposedly lethal injections by Seeling as the nominated medical representative, but these appeared not to have taken full effect, so the men were finished off by strangulation or having their chests stamped on.[9] One man did not react at all to the injection, so he was taken to Gestapo HQ where he was shot by Hoffman.[10]

Another trial related to the *Freshman* disaster opened in Hamburg on 28 May 1946,[11] the defendants being General Karl Maria von Behren who had been *Ortskcommandantur* at Stavanger[12] in November 1942, together with von Falkenhorst himself. The accusation was one of war crimes – that Behren had been instrumental in the murder of fourteen British POWs (at the time unidentified) at Slettebo Camp. This was located near Egersund and the survivors from 'A' Apple's glider were transported there. At this camp they were interrogated, then all marched up to some woodland to the rear of the huts where, one by one, they were shot dead. Every man, waiting his turn, would know what had befallen his comrades and that the same inexorable fate was his to share.

Under cover of darkness, the dead were thrown into the back of trucks and then dumped in a shallow mass grave. A witness at the trial, a Norwegian labourer who was one of those conscripted to carry out these grim chores, stated he believed that some of the victims were still showing signs of vitality and continued to bleed even as they were thrown into the shallow pit.

A couple of months later and it was von Falkenhorst's turn in the dock.[13] The general, no longer a strutting Prussian Junker but still, even if he'd been no zealous fan of Hitler's, a man never refused an order, was up on nine counts. All of these were allegations of war crimes with the third and fourth indictments relating to the *Freshman* atrocities. Charge number three concerned the murder of the fourteen PoWs killed at Slettebo Camp and charge number four accused him of handing over nine PoWs, alleged saboteurs, to the SD. His culpability arose not so much from direct personal complicity but from the fact he was C-in-C.

In November 1942 he had instructed, indeed in terms of criminal liability, incited, men under his command to carry out the infamous order. Even if he had originally protested to Field Marshal Keitel, he had acquiesced and just so the message was clear, he'd added his own addendum reminding his troops that if they did capture any Commando types, they were to be interrogated then executed within 24 hours. This was about as unequivocal as it gets. And it came with a final reminder that the Germans involved should maintain strict secrecy and ensure there weren't any inconvenient Norwegian witnesses present. This clearly implies he knew perfectly well there could be repercussions, and even if the occupiers were only obeying orders, there was no need to broadcast the fact. The General did concede that the shooting of PoWs by the Wehrmacht was incorrect, only insofar as they should have been handed over to the SD. In which case, summary execution could almost be considered a kindness.

Feldwebel Kurt Hagedorn provided testimony. He had been at Slettebo at the time of the murders and had witnessed the doomed survivors being individually taken to their deaths. He confirmed they all appeared to be wearing uniform. The court found the defendant guilty on seven of the nine counts and the 'guilty' verdict extended to both accusations relating to Operation *Freshman*.[14]

In July 1945, the dead British soldiers were properly re-interred at Eiganes Churchyard with all due ceremony. Throughout the war loyal Norwegians had marked and tended the mass grave until the oppressor was finally vanquished and those murdered could at least receive a decent burial.

Operation *Freshman* was a disaster driven by necessity but untainted with practicality. This was Britain's first glider-borne operation, still an unknown quantity and the Hardanger Plateau was a hard target. There has to be a question mark over whether those who died were a noble sacrifice or culpable waste. The loss of so many brave young men will always be terrible, but it wasn't the end of the fight. Other brave young men were already on the ground. A handful of others would soon join them and what they achieved would become the stuff of legend.

This fight wasn't over, it was only just beginning.

6

SWALLOW

> 'Bombing is impossible. The factory is deeply embedded in a gorge. The only way across that gorge is a narrow bridge, 75-feet long. Two men with machine guns could hold off a battalion.'
>
> *Heroes of Telemark*

'The November 20th wireless message about the glider disaster was a hard blow. It was sad and bitter, especially as the weather in our part of Norway improved during the following days.'[1] This was quite an understatement. To have endured everything Team *Grouse* had gone through, not just the mission but all their preparation, training and the fact their nerves were screwed up to breaking point at the decisive moment when they finally got to strike a blow, only to have the chance so brutally snatched away, was not merely sad. Add to this the fact that that those men who had been depending on them had all gone to their deaths. Not their fault, but the burden of loss, disappointment and guilt must have come down like a leaden pall. This surely was their darkest hour. The fact they would rise beyond this, endure so much more and finally, through sheer grit and determination, achieve their finest hour, really says it all.

Helberg picks up the story:

> Only one of the planes had returned. One glider had been dropped over southern Norway, and the other plane with a

glider was missing. We had nothing else to do than to wait for further orders. We did not feel very safe in our cabin. The German camp at the Mos Lake Dam was only 5 kilometres away. Consequently, we went northwest again, toward Saura, between Mogen and Kalhovd. There was a little hunting cabin there that was not marked on the map, and it became our home for three months.

More Hard Times This period offered us a fair share of difficulties. At the outset the Germans were especially worrisome. One day when I was down at the Mos Lake to get the latest news from our contact, Torstein, I was told that he had just been arrested. There had been large raids in Rauland, Mosstranda and Rjukan, and many people were arrested.[2]

Almost as bad as the failure and losses of Operation *Freshman* was its inevitable stirring of the hornets' nest. The enemy was fully aware of their intended target and would have quickly worked out that the raiders must have had ground support:

A German radio-direction-finding car had been observed at Rjukan, and German patrols had headed for the mountains. Every house in Rjukan and the Vestfjord Valley was searched, and the Germans asked everywhere if people had seen Englishmen in the area. The reason for the raids was obvious. It was a known fact that the British from one of the gliders had been captured and killed by the Germans. In the glider, the Germans had found maps and information showing that Vemork was their target. The Germans also almost certainly deduced that a radio transmitter in the area had been in contact with the British. Their suspicion fell on Einar Skinnarland, Torstein's brother. But because he was not at home when the Germans came to arrest him, they took Torstein instead.

The Germans were closer to their target than they realized. Einar belonged to our own group. He had, however, left England as early as April 1942 to furnish the Allies with information on German activities in the Rjukan region. We had not been in contact with Einar yet, but he joined us later in the winter. After we lost our contact with Torstein, our

> food situation became difficult again. For two weeks, we lived on oats, some margarine, sugar and Iceland moss. We dug the moss up from under the snow and boiled it. It hardly gave us any nourishment, but we imagined it did, and that helped a little.
>
> On this diet, our strength waned quickly, and we were susceptible to diseases. At one time, we were all in bed, except the chief. I believe he managed to keep healthy because he spent much time thinking about reindeer steaks … Meanwhile, the radio connection with England was maintained. We were told a new party would be sent, but only one plane would be used, and that it would deliver no more than six men, all from the Linge Company. The fellows would parachute to our quarters at Saura.[3]

Baker Street was equally despondent. Neither Tronstad nor Wilson had had much enthusiasm for the glider-borne operation, preferring, from the outset, Norwegian Commandos secretly infiltrated, whose local knowledge would be their prime asset. Despite the extent of this awful tragedy, *Grouse* were still in place. There was hope. Jomar Brun, an invaluable resource, had now arrived safely in the UK, his fully up-to-date information on the actual layout of the plant was a godsend. Wilson contacted Combined Ops by phone and offered to take over responsibility for the mission from now on. 'Thank God, please do,' was the response. This was to be, despite prevailing despondency, a real game-changer.

Meanwhile, on the remote tundra of the Hardanger, the Commandos' precarious existence continued, monotony with a shrieking chorus of winter storms, and the pang of empty bellies. Helberg recounts:

> Shortly before Christmas, there was a lot of rain and fog. Such troublesome weather suited us admirably. The Germans got tired of their patrolling and the wild reindeer came down from the western areas of the plateau. From then on, our chief was out daily to bag an animal, but conditions were not

good for hunting. The fog was thick in the mountains, and he was so weak that he could only get to the nearest peak. But on Christmas Eve we received a welcome gift, clear and fine weather. Our patient leader went out again; this time he returned with reindeer meat. For two months, we ate only reindeer meat, as many as 11 animals.

To prevent scurvy, we ate the stomach and its contents from every animal. It tasted quite good, when mixed with blood and fat. Tough sledding. There was miserable weather in the mountains that winter, and we could not go hunting often. For weeks, the big storms would rage, and it was almost impossible for us to go outside ... Polar disease is the name of a disease attacking people who live in isolation under difficult conditions. In a short time, the best friends may become the most bitter enemies. Even the most cheerful people may collapse. The fact that we did not fall victim to diseases was largely due to our leader's creative ability. We had no light, which meant long evenings.

Idleness is the cause of polar disease to a great extent. To prevent this, our chief started a study circle. Each evening, one of us gave a lecture. The theme did not matter; the main thing was our thoughts were removed as much as possible from our daily existence. From time to time, I went for a walk down to Rjukan or to the Mos Lake. We had a couple of contacts who furnished us with information on the movements of the Germans in the Vestfjord Valley. Meanwhile, the situation at Vemork had become less favourable.

It looked like the Germans were prepared for an attack, as the number of guards at the plant increased steadily. The Austrian guards were replaced by Germans; there were double guards where there had been single ones, and mines were deployed. Mounted floodlight facilities were installed so that the whole factory area could be lighted up as bright as day in case of alarm. We sent reports on these developments to England. The transmitter demanded much electricity, so we made strenuous trips with the battery to and from the charger. And I had to make many return trips.[4]

Wilson wasn't idle either. He and Tronstad were singing off the same hymn sheet. That very morning, he contacted Kompani Linge in Scotland to prep Ronneberg to be ready for a hazardous mission – in the certain knowledge the 22-year-old officer wouldn't need to think twice – and that he should choose another five to go with him. Next, he briefed Gubbins, his chief who was understandably alarmed, risking yet more lives when so many had already been lost. Wilson was adamant. This was how the job should be done and these were the right men to do it. The War Cabinet concurred. Vemork was too great a prize for anyone to entertain lingering qualms. If the Germans thought their barbaric treatment of the *Freshman* survivors had scotched the snake, they couldn't have been more wrong.

Next winter's day in the chilled greyness of a highland dawn, Tronstad and Wilson were on the overnight train north, ready to brief Ronneberg in person. To their great credit they didn't attempt to gild the lily. Ronneberg was told why his team were going, because the previous mission had failed catastrophically and, in all probability, the Nazis had now murdered their prisoners. Failure was an automatic death sentence. Being in uniform wouldn't save you and of course the enemy was now alerted and would step up security in and around the plant. On paper, it was mission impossible.

Wilson knew how to judge his men, many who had signed on initially had been transferred to regular forces and many simply hadn't made the grade, had failed to demonstrate those characteristics essential for a successful agent. The 'gangster' types so seemingly beloved of General von Falkenhorst just weren't right. Thuggery might be needed, but not thugs. What was required 'could only be provided by men of character who were prepared, to adapt themselves and their views – even their orders at times – to other people and other considerations, once they saw that change was necessary. Commonsense and adaptability are the two main virtues required in anyone who is to work underground, assuming a deep and broad sense of loyalty, which is the basic essential.'[5]

Like Poulsson before him, Ronneberg didn't so much as flicker. Moreover, he already had his mental list of companions prepared.

Haukelid, now fully recovered from the negligent discharge and since commissioned, was an obvious choice: '... twenty-nine years old, a powerful man with a chest built so high with sinew and muscle that it looked almost bulletproof, a hunter's love of the outdoors, a philosophical turn of mind – a kind of Henry David Thoreau with brawn. His unidentical twin sister was the internationally known actress Sigrid Gurie, who shared his anti-Nazi views with equal fervour.'[6]

Keen as he was to get into the fight, Haukelid's time thus far with SOE hadn't been all blood, sweat toil and tears – far from it:

> We were sent to STS no. 61, an establishment in a country house in the neighbourhood of Cambridge. It was a station for people who were going to Europe on secret errands, and who had to wait for planes. The place was very closely guarded. A number of service women kept the house in order, cooked the meals and gave the men some social life. They belonged to the special section called 'Fannies'.[7] These girls did an uncommonly good job ... And when the commandant suggested it they were always willing to come to Cambridge in the evening for a little party ... but if we asked the fannies about our comrades who had gone out before us, they became dumb and knew nothing.[8]

Happily, and obligingly, the Fannies usually had their own cars, 'and very fine ones'. The Commandos dined well and at the taxpayers' expense. After all, such opportunities might become very rare. Very soon they did. 'Joachim Ronneberg was a conscientious chief, and saw we were home early, got plenty of sleep, and were always ready for an operation at any given moment.'[9] As Haukelid himself ruminated, this war was unlike any that had gone before; mobilisation for SOE meant a 30-minute warning, not weeks of ponderous manoeuvre; one evening you could be swilling fizz in a club on HMG's account, the next you could be chewing pemmican on the Hardanger Vidda.

Throughout, however charming these young women might appear, they, too, were on a mission, to keep their young men safely on the leash; even if it was quite a long one and to cosset them from mishap. They were being kept like Lord Raglan's light cavalry 'in a hatbox' – just so they'd be ready for their own Balaklava when it came.

SOE didn't realise or chose not to realise the unorthodox uses to which Haukelid and the others sometimes applied their extensive training: 'All the same we kicked over the traces now and then. The country around the estate was full of beautiful pheasants, which made excellent eating. Occasionally, we would crawl out through the barbed wire fence, and shoot the birds with pistols in the fields and copses ... Sometimes the neighbouring farmers complained to the commandant that strange bandits were shooting on their land, but the commandant was not aware that he had any visitors in the house at all. That people who did not exist could shoot pheasants was, he said, impossible.'[10]

Sergeant Frederik Keyser was next up, 'slender and wiry', about as far from the musclebound type as you can get but with a core of steel. Hans Storhaug was quite similar, an 'unheroic-looking man' (known to Haukelid and his other comrades as 'The Chicken', having once been nabbed red-handed by a Scottish ghillie plucking one of the latter's pheasants). The outraged gamekeeper had confirmed the thief 'had a ruddy great beak and looked just like a chicken'. Happily, this description was insufficient for a conviction.

Kasper Idland, another quite innocuous looking fellow, was fiercely motivated by a blazing disdain for the Nazis occupying his home. He carefully hid a deep personal secret – he was an awful skier! His SOE trainers had assumed all good Norwegians were naturals and he saw no need to correct them. Birger Stromsheim was last but by no means least, a man of infinite strength and patience and the perfect foil to Ronneberg's D'Artagnan. These were all very remarkable young men. And, of course, they had no idea of the crucial importance of the mission.

Jomar Brun proved an invaluable source. The six chosen men were moved to STS no. 17 where, under Brun's direction, a full-scale mock-up of the basement and its eighteen high concentration cells had been constructed. The team relentlessly practised entering the space and laying and setting their charges, until it became second nature. They spent hours familiarizing themselves with the devices they'd be using, building actual charges (one for each of the cells), along with studying the overall layout of the plant itself and surrounding ground. Tronstad was their ringmaster, he took onboard all their questions and any he couldn't answer on the spot he'd note down and research a proper response (the team didn't know about Brun, Tronstad's 'deep throat').

It was under Brun's instruction that the high concentration room had been increased in size. He knew everything there was to know about the cells, how they were fabricated and positioned, how explosives might be laid to inflict maximum damage. He was aware of where the guards post stood in the factory yard. All pure gold. He could tell them of the one entrance that was left unlocked – silent entry, the rest were barred and being of steel construction, would need to be blown, overture to a firefight. Why was one point of entry left accessible? This wasn't a doorway as such but a large cable duct, running from the exterior into the basement target area, a gift from the old Norse Gods.

Brun had firsthand knowledge of it. Prior to his flight, he'd had to crawl along the duct to investigate a leak from one of the pipes taking caustic into the basement, or at least down to a manhole which lay atop one of the cellar rooms, directly adjacent to the high-concentration vault. 'Very few people even at Vemork, knew about this entrance.' For all von Falkenhorst was surrounding his crown jewels with a ring of iron, there was this one chink in his armour. It might just be enough. If any of the *Gunnerside* men were wondering just how Tronstad, so long an exile, knew all this, they knew better than to ask. In fact, the less they knew, the less could be extracted if they were captured. Nobody had any illusions about the Gestapo's persuasive capacity, that was why they carried suicide pills. (This final resort was also dubbed the cyanide pill, kill-pill or L (for lethal) pill. About the size of a pea, it contained high concentrate potassium cyanide sealed in a thin-walled glass ampoule, coated in protective brown rubber to prevent accidental breakage. Once crushed in the mouth, death would occur in a minute or so.)

Ronneberg and his merry men made ready as 1942, a year of mixed fortunes for the Allies, ended. The race for the Bomb continued. The Anglo-American effort had already outstripped the Axis, but London and Washington remained understandably paranoid over just how close they might be. General Rediess had already signalled to Berlin that Vemork was still a high priority target. On 4 December the Germans, under the cover of a dummy air raid warning, blitzed Rjukan with two full companies of security troops who methodically searched every dwelling. It took them 15 hours and was enacted with typical Nazi thoroughness. They still came out empty-handed.

Von Falkenhorst returned to Vemork along with Terboven. The general's apprehension was extreme. Though a failure, *Freshman* served to show what to lengths the British were prepared to go. A wave or arrests coincided with this inspection. One of those netted, Thorstein Skinnarland, would sit out the rest of the war in an internment camp. The Nazi C in C would have been even more jittery if he'd any idea just how close the *Grouse* Team (now renamed *Swallow*) actually were. In fact, their current base was only 2½ miles (four kilometres) distant from the garrison stationed on the Mosvatn Dam. Too close for comfort, time to move on.

Withdrawing in a north westerly direction deeper into the wilderness of the Hardanger, they paused for a few days in a hut at Grasdalen before pressing on to a small shooting lodge – one that Poulsson and his cousin had erected in the halcyon days before invasion, hence the team referred to this as 'The Cousin's Cabin'. It was remote, not shown on any map, and located in the Saure district around 18 miles (29 kilometres) north west of Rjukan. Inside was a visitor's book and Poulsson found recorded there his own last visit, in the summer of 1939 ... echoes of a tranquillity now lost, as distant as another century. In those vanished days of a fine summer when the Hardanger presents an altogether kinder face, how could he have envisaged under what circumstances he might return? Now, even if they were not fully informed, they were stalking something which represented the ultimate blasphemy, a weapon which could wipe all such tranquillity from the face of the earth forever. The innocence of those warm days was as ancient as the pyramids. Here was a new and terrible world.

London had confirmed that *Grouse/Swallow* could expect another team of saboteurs to be parachuted in, a maximum of six Commandos to be dropped near the Cousin's Cabin. Baker Street couldn't say exactly when. Nor could it suggest – despite the care being lavished on the selection and preparation of Ronneberg's group – how Poulsson and his fellows were supposed to survive in the interim. It was assumed that they'd just keep muddling through, as before. The plain fact was Baker Street didn't really understand the Vidda at all.

Torstein Skinnarland had been their life support mechanism, but he was gone. Their miserable diet now consisted of oatmeal, a dash of margarine and reindeer moss. This is a form of lichen that grows on rocks and in light soils, on average the coating is around an inch (25 mm) in depth. It doesn't look nutritious or appetizing. Yet, as Poulsson, the mountaineer confirmed, reindeer could survive through the winter on this stuff. Like native *tarandri,* the Commandos burrowed through the snow to harvest the moss, then boiled it up with oatmeal. Poulsson waxed eloquent as to the rich nutrition this desperate diet offered.

Poulsson may have oversold the benefits of this lichen because both Helberg and Kjelstrup developed oedema, caused by a build-up of excess fluid in the body, causing affected tissues to swell – most often in the legs and feet. Such a painful and debilitating condition further weakened the sufferers. In both cases their weight rocketed, and they suffered from disturbed sleep patterns as they were up to urinate half a dozen times at least every freezing night. Just about everyone was ill in one way or another – though Poulsson, swearing by his mossy diet and consuming more than anybody, did fare best! He tried to raise flagging spirits by summoning up the intoxicating aroma of reindeer steaks. He knew the animals wintered on the Hardanger Plateau and their flesh would prove the team's salvation. If only.

On 10 December, Haugland discovered a well-oiled Krag rifle and boxes of ammo stashed inside one of those shielings dotted around the small lakes. Poulsson now had tools for the job and skied off every morning in search of game, scouring the barren waste, seeking higher and higher up mangled scree slopes and boulder-strewn summits. Mist, fog, wind and despair were his inevitable rewards, the bitter disappointment with which every day he slunk back to pinched and sickly faces. Illness produces lethargy, men became faint and dizzy, easily exhausted, lacking even the essential resolution to gather firewood, as vital as food to their survival. It was their very lowest ebb, not the white heat of battle but the lonely, ever debilitating fight against nature. A fight that nature must always win.

But then, on 23 December, those shots that spelt survival. Poulsson killed their first reindeer. He'd also wounded another and with the warm lifeblood of the dead animal spreading a reviving heat through his veins, he went off to search for the other one he had hit. There was a blood trail but minor and intermittent. He found no stiffening carcass and judged the injured beast would have survived to rejoin the herd. Still, he had the main prize.

Like the accomplished woodsman he was, Poulsson set about preparing the kill. First into his rucksack went the head and tongue, graded high on the list of cuts. Next came the ribs, brisket and then internal organs. Like Native Americans, whose culture he had studied, nothing was left to waste, leg bones were included for their rich and nourishing marrow. He chewed on sections of fat to provide energy, sucking on marrow from smaller cuts to fuel his tired body: 'In these small bones, which help give reindeer their extraordinary spring and elevation, the marrow is liquid and almost as clear as water.' In this instance literally the water of life, its effects were immediately energizing. The dead animal's intestines were also harvested, being rich in vitamins and minerals. In all, the catch weighed in at about 50 pounds (23 kilos).

Covering the rest of the haul with the dead animal's pelt (this wouldn't be wasted either, used to insulate their cabin), he set off for the return trip, heavy sack bulging, frozen pail of blood bouncing:

> I felt tired under the weight of the rucksack, but happy, too, as I poled along the same way I had come. Then, as I sped down the slope towards Angelbutjonn Lake, I came across the large herd of reindeer I had encountered that morning. They were relying on the ice in almost the exact same place and didn't notice me till I was right by shore. Suddenly, the entire herd, in a compact mass, started off. There was a thunder of hooves, and a shrill, creaking sound from the rubbing together of their horns. In a second, they were gone, leaving only the trampled snow behind them. But I knew they'd be back, along with other herds migrating from the west.[11]

Famine was ended and vigour could be restored. The Hardanger was finished collaborating with Germany.

Well perhaps not quite finished, as Helberg recorded:

> Often, I was unable to force my way through the storms. Fair weather finally returned one day at the end of February. It was real 'drop weather', with moonlight and quiet, but 34 degrees below centigrade. Over the radio, we were told that the men would come that night, which meant that we would stand outside with the lights. We stood there in the cold until 05.00, but nobody came. The next day we were told that the plane had not been able to find us and that it had dropped the men on the High Plateau, most likely 50 kilometres north of our position.
>
> At the same time, we were hit by a terrible storm, the worst we had experienced. We had to stay in the cabin for a week. '*Gunnerside*' was the code name of the new group. They were given a rough reception: the big storm struck even before they had collected all their equipment. They were lucky, however, as there was a large cabin right by the landing site. On the first day of good weather, they headed south, and in the evening, we saw them coming from the west. It was fun to talk with

> friends who had news to tell. We had lived for five months alone in the mountains, and the change was welcome.[12]

Before this happy meeting, in the meantime for *Swallow*, survival, if never guaranteed, was at least sustainable. Fresh meat meant renewed vigour, fuelling the daily round of chores and foraging for kindling, maintaining radio contact and waiting, just waiting. As Poulsson recollected:

> Shivering with cold, I crawl out of my sleeping bag, stick my feet in my moccasins, light the kerosene lamp, and drop down on the stool by the stove. It is 07.30 and today it's my turn to fix breakfast ... From long and bitter experience, we have learnt that as long as there are storm clouds over the Vrasjo Mountains, there will be no attempt to land the *Gunnerside* men and their supplies by parachute. The flying weather is probably even worse farther west. We prepare ourselves for our day's work. Claus is going reindeer hunting again. Arne is going out to look for wood. Einar is to help Knut with the radio transmission. Myself I start the day's work as a cook, and soon it's time to think of dinner. A new day has begun at the Cousin's cabin, a day like so many others.[13]

In Scotland, team *Gunnerside* was enjoying or rather enduring its own routine. Each day the men would be psyched up and ready, each day the mission would be aborted due to adverse weather. That finally changed on 23 January. Marching with full packs in white camo suits, they debouched at the lonely airstrip, where nobody looked at them, no eye contact, select personnel were not in the habit of engaging or asking questions, no idle banter. Tronstad, the good shepherd, recently made up to full colonel, was there to wish them *bon voyage*, though there wasn't much *bon* about it. Their bravery of these half-dozen heroes, and whilst the word today is ridiculously overworked it fits, is highlighted by the fact they still had no real appreciation of the crucial nature of their mission or indeed the full tally of risks involved. Leif Tronstad could only assure them that their actions would be vital to the Allied cause.

By way of a training postscript (or codicil), each was issued with a suicide pill. Tronstad had been completely level with them – after the catastrophe of Operation *Freshman*, they knew better than to expect humane treatment or quarter if they were captured.[14] Finally, they were secured in the belly of a chunky Halifax and rising into the night sky.

Now cruising at 10,000 feet, they crossed the coast of their motherland. Pulses quickened. Was it finally to happen now, after all these frustrations? Stomach muscles tightened; a new tension flooded. But nature hadn't finished toying with them. As they approached the drop zone, cloud closed in again. *Swallow* had marked the L of the landing site and Poulsson, having detected the roar of aero engines, flashed white light from his torch, to contrast with the red.

He'd been here before and the steady thrum of hope receded westwards into the still abyss of freezing dark, swallowed behind serrated mountains. He kept it up for another half an hour, but the Halifax did not return. It's impossible for us to put ourselves in Poulsson's place, the sheer frustration and encroaching despair must have been seismic, but he steeled himself to just keep going. Meanwhile, the pilot, unable to detect the lights, decided to abort and return to Scotland. Ronneberg and his team experienced their full measure of frustration too.

Tronstad, who understood only too well the potential consequences of endless delays, was under pressure from the very top of the command chain. General Groves, increasingly worried the Germans might yet win the race, was leaning on Eisenhower, now appointed as supreme allied commander in the west, to neutralise this threat at all costs. If Commandos couldn't do it, then it was time for the bludgeon of the heavy bombers – Tronstad's recurring nightmare.

Ronneberg had decided that if on their next attempt they still couldn't detect the drop zone, they'd jump blind, just get onto the ground, and then make their way to join Poulsson. Whilst this was very far from ideal, it did avoid the dreadful anti-climax of coming so close only to turn around. In the small hours of 17 February, the next full moon, they again flew into the blackness and just jumped. All came down safely with their gear. 'Where are we?' Ronneberg demanded of Haukelid (as nobody knew the Hardanger better). 'Not a clue.'

They were in fact some 30 miles (50 kilometres) north west of the Cousin's Cabin, close by the frozen waters of Lake Skyrken. The RAF had played safe by avoiding flying over either Rjukan or Lake Mos and alerting the enemy. Secrecy was maintained, but it was also a long way to traverse in winter conditions. First, they had to gather in all their kit while Storhaug, by far the most accomplished skier, set off to recce suitable havens. Good news, there was a serviceable cabin under a mile (1.6 kilometres) distant. Their equipment was located and stashed in a well concealed laager, marked for future collection. Their new home soon opened with a well delivered axe blow and they were in from the howling, knife-edged wind. Obligingly, its owners had left a neatly stacked pile of wood by the stove. *Gunnerside* was in business.

Perfect skiing next day, surely a good omen for the team, surplus supplies safely stowed, carrying just what they needed for the mission. But this was the Hardanger which doesn't do favours. Soon they were being driven on by buffeting wind, propelling them like a giant's hand. Still, the blast would surely speed them towards their friends. Nonetheless, Ronneberg was worried. Worry is the function of command, and his present angst was dictated by the fact their balletic progress should have been glissading over a frozen lake. But they weren't, bushes don't grow on water, and he began to succumb to the horrible suspicion they were headed in the wrong direction. Despite their extreme fitness, the weather conditions were sapping energy reserves, muscles were leaden, neck glands swollen, and relentless snow kept falling. For all their training, Scotland wasn't the Vidda.

He decided they must double back to the relative security of their last night's hut. All their tracks had been swallowed up, the demonic pitch of the wind's demented wail, its devilish blast now in their faces. Mercifully, they virtually bumped into the bothy and crowded inside its shelter, fearful, goggle-eyed snowmen. Faces frozen in weird rictus, it took time to get their chilled features to obey instruction. As they struggled to recover breath and composure, the tempest did not forget them, it pounded and shook the flimsy structure with a giant's wrath.

Where the devil were they? One slice of good luck was a map on the cabin wall and Ronneberg, by drawing a series of circles from where they should have been, finally identified a thumbprint smear on the paper. Ah ha! Next, they forced open the hut's inner door into a second smaller room where they providentially discovered a visitor's book. This identified their true location, so they decided to recuperate and dry out while the storm fumed, then pick up skiing as soon as it cleared. It didn't. This was the mother of hurricanes, stirred and swirled by the restless streams of air blasting across the fissured surface of the plateau, whipped into cyclonic fury, spilling out accumulated moisture as driven snow, flakes that stung like grapeshot.

God alone knew how the cabin stayed intact. At any minute they dreaded the thump of a titan's contemptuous hand that would send the whole thing and them inside it spinning away in splattering fragments. They were helpless in the face of such elemental rage. Nothing in their training could help them, their carefully selected and hoarded kit of no recourse. Even the mission seemed like a fantasy, so wholly was their consciousness dominated, absorbed and confounded.

One of the chimney braces snapped under the onslaught. Ronneberg volunteered to fix the problem. Now outside and clinging precariously, the tempest seemed to focus wholly on him, thick white flakes whipped into a mesmerising dance that obliterated everything. Only a supreme effort of will kept him to his task, he clawed for balance while locating the damaged end of the bracket, clinging to the one still in place. Like a playful demon, the storm buffeted him one way and then the next, offered a few seconds clarity them blotted out everything: 'The wind begun to rock the grass / With threatening tunes and low / He flung a menace at the earth / A menace at the sky.'[15]

One capricious gust, that bit more vicious even than the rest, suddenly whipped him from his precarious perch and hurled him yards off, providentially into a snow pile. As he struggled to rise, he was bowled over and over again like a child's toy. Blind and disorientated, he staggered, hands held out, till he touched the shieling's wall. The gods weren't done, not yet. All through that night and next day nature kept up the pressure. Though all were Norwegian, the Commandos had never witnessed snowfall like this. It ground them down, colds turned into fever, movement an enervating exercise in willpower. This storm, its sustained fury battering the senses, induced at times an almost dream-like quality, shut up in the fragile womb of their borrowed shieling. On day five it finally subsided, and the skies began to clear. It left a world of dazzling virgin whiteness which, thrashed by the relentless blasts, had been sculpted into an unrecognisable landscape of huge white dunes, surrounded by a jagged confusion of icebound peaks.

Ronneberg fretted that *Swallow* might have written them off as they'd been unable to establish any kind of radio contact. First, they needed fresh rations from their stash, which in this weirdly transformed universe, took some finding. Their quest was at least successful.

Time to set off again but, as they loaded up, about to finally quit their refuge, they were stunned to see a couple of hundred metres away and dragging a sled, a solitary figure. He must have seemed like an apparition from mythology. Who was this lone skier? He hadn't detected their presence and continued towards the hut, an everyday activity, unconcerned and clearly an experienced skier. He was soon disabused. Taking fright at the fresh ski tracks, his confusion turned to outright terror at the sight of bearded ruffians bristling with levelled weapons.

Dissolving in fright, the man grovelled in the fresh, white snow. Instinctively, he must have realised this could mean the end for him. Ronneberg questioned the fellow closely. His name was Kristian Kristiansen, from Uvdal, a small town on the north-east flank of the plateau. The contents of his wallet bore this out. He was there to hunt reindeer for his black market sideline. Was he a member of Quisling's faction? Now this was tricky. Kristiansen would struggle to tell allied battledress from Axis, especially in snow gear. All logic would dictate these must surely be Germans. He replied evasively, not a member as such but a tacit supporter. Wrong answer.

A hurried discussion was held. What to do with this bloke who'd just shot himself in the foot by admitting Quisling sympathies? SOE would have had no qualms, nor did some of the Commandos. Kasper Idland was adamant Kristiansen must die and he'd do the job himself to salve anyone else's more tender conscience. Ronneberg was undecided, killing an unarmed man, a civilian and fellow countryman was repugnant. He questioned Kristiansen again on his alleged sympathies. The man was wary, convinced these were Nazis and whatever he said could only incriminate him further. At length, they tried to convince him they were in fact Norwegian, but he was by now sure they were just trying to catch him out. It wasn't until someone made a chance comment in the local dialect about King Haakon that the penny finally dropped. Suddenly he was indeed a loyal patriot!

Ronneberg was taking a huge risk, Kristiansen was now gabbling excitedly in dialect, all aquiver to serve his King. By way of insurance, he was made to write out a full list of his illicit activities, naming his customers and sign this as an assurance. They'd have to leave him his rifle and skis. Ronneberg knew he was putting both the operation and the lives of the whole combined party at hazard. If the man talked, they were blown, simple as that. He decided on a temporary compromise, using Kristiansen as a scout/hostage, putting his local knowledge to good use whilst keeping an eye on him. There were still mutterings and objectively, the man should have just been silenced, but Ronneberg had come home to fight tyranny, not subscribe to it.

The Uvdal black marketeer proved an unexpected asset. He knew every inch of ground and moved like a maestro on skis, gliding over ground and always finding the easiest contours. With him as their guide, the trip seemed so much less onerous. Setting off in the predawn, they paused at a cabin their scout knew well; according to him it belonged to his brother-in-law. Breakfast, an expenditure of scarce resources, including real coffee marked a new note of optimism. Reindeer were much in evidence, frustrating for Kristiansen who did not have permission to stalk. He was nonetheless a true mountain man, his eyes keen as a hawk's and it was he who spotted two skiers approaching. His terrors returned with interest when the Commandos instinctively went to ground around him. This war was suddenly very real. Deprived of his skis lest he be tempted to make a run for it, he burrowed frantically into the snow.

Ronneberg signalled to the sharp-eyed Haukelid to scan the visitors with his glasses to see if he could recognise any of Poulsson's group: 'But I could not say who it was. He was packed in so many clothes that it was hard to get a proper view of him, while a huge beard made the face face unrecognizable. The situation was not improved when we saw another ski-runner, coming up the valley in the track of the first man and a couple of hundred yards behind him. Could the second man be Arne [Kjelstrup]? The only thing that made me doubt was that he should have such a red beard!'[16]

Team *Grouse* was lying on high ground overlooking Grasdal Lake. These newcomers went over toward the lake, topping a small rise, still keeping distance and clearly scanning the plateau above. Ronneberg sent Haukelid forward to investigate and while he was pretty sure these were their comrades; he still kept a pistol securely hidden beneath his parka.

'Ten minutes later I was standing on top of the hillock. In the soft snow I had come right up behind the men unnoticed. I stopped about 15 yards from them, leaned on my sticks and looked at Arne and Claus. It was small wonder that I had not noticed them at once. They were wearing what in an English shop would be advertised as the last word in sports clothes; but these were so ragged and disgusting that the boys looked like the worst tramps imaginable. Behind their beards their faces were wan and thin, and their shoulders were bent'.[17] Those months spent living on the Vidda had taken their toll.

Both Kjelstrup and Helberg were busy squabbling over who should get first dibs on their single telescope. Haukelid, still unnoticed, coughed loudly. This unleashed a few seconds of tension as both parties drew pistols, best Wyatt Earp style. And then, that wondrous moment of recognition. 'There was back slapping and much strong hearty cursing. We ran down to the lake waving to the others ... we expressed our feelings as strongly as young Norwegians can.'[18]

As the hallowed lore of hospitality dictated, *Gunnerside* broke open their ration packs to shower bounty on their worn-out comrades whose bad physical shape was a testament to the hardships they'd endured. Crisp breads, powdered milk, dried fruit, raisins, chocolate and bliss, tobacco; manna direct from heaven (via England) for men so long deprived.

This still left the group problem of what to do with their conscripted guide, now recovered from his funk as clearly there'd be no gunfight. Idland still wanted him dead but the rest demurred. He was made to accompany them back to the Cousin's Cabin. Kristiansen was forcibly reminded that he was best off keeping quiet. The list of his activities and contacts would be very embarrassing and the mere possession of a rifle, a death sentence. By way of incentive, he was loaded with the gift of a week's rations, cash and that rarest of luxuries, chocolate. He was finally cautioned to keep away from Uvdal for a further three days, during which time he was fully at liberty to resume hunting. All in all, he was a very lucky man. Willie Fairbairn would have been horrified he was still breathing.

7

GUNNERSIDE

'I can get inside that factory with ten men. Ten men, and blow it sky-high.'

Heroes of Telemark

These are your sons –
Coal, steel and iron have tinged their blood,
Fresh, bright blood now mingling with scarred veins,
Bringing new strong vigour
And young, gay laughter to the northward pour
Of armies where the Elder and the Younger sons
Are side by side.

E. Yates, 'These are Your Sons'

Men who have known what hunger really means can be excused from taking time out to celebrate bounty when it finally arrives: 'Their eyes shone like little children's on Christmas Eve.' Haukelid could understand: 'As soon as the flood of talk had in some degree abated, we offered each other the good things we had in the way of food. To offset our imported delicacies, the *Grouse* lads had reindeer meat of every variety. As they put it, we could have the meat either roast or boiled, our boiled or roast according to our taste and fancy. Jens [Poulsson] who remembered my shooting expeditions in the mountains in former times, knew that I appreciated the best part of the animal, and we silently divided a couple of marrow bones. I knew from experience how dangerous it is to eat too much reindeer meat when the stomach is unaccustomed to it. I ate too much all the same and had the same stomach trouble which the *Grouse* boys said had afflicted them!'[1]

Reunion was joyful, but nobody was there to enjoy themselves. They might not understand fully the absolutely vital nature of their mission, but they knew it had to get done:

> I had brought a special message to *Grouse* from the Supreme Command of the Armed Forces [Norway]. While the *Gunnerside* party were out skiing I read aloud the order about the organizing of future work in the province of Telemark. The message was printed on rice paper so it could be eaten in case the bearer was taken prisoner. 'Is that food?' Knut asked as I crumpled this paper up and was about to throw it into the fire. "'Quite digestible.' 'Well, we don't throw away food here at Svensbu.' He took the paper and began to chew. It may have been a bit tough but the lads at Svensbu were not spoiled.[2]

Everyone knew time was of the essence: 'The day after we arrived at Svensbu the question of the attack on the heavy water establishment at Vemork was taken up. The boys at Svensbu had continually been obtaining fresh information from Vemork. Now we all sat together and wrote on small scraps of paper all the questions to which we wanted answers. The scraps of paper were laid in a heap and the questions sorted. Finally, Claus [Helberg] was sent off to Rjukan to get the very latest information. He had with him a list of over 30 questions.'[3]

Helberg and the rest of *Grouse* were very keen to get into action, only now was their incredible fortitude and stamina about to be fully deployed:

> We immediately began to plan for the attack. We had to figure out how to get into the factory area and from there into the factory itself, and then how to get back alive. We had two alternatives to get to our target. One was the bridge across the gorge by Vemork, but it was guarded.

> Liquidating the guard would make noise and perhaps reduce our small force. Another way was to try to cross the gorge under the bridge, and from there move unseen into the factory. We knew that the Germans believed that no one would try it that way. Our contact from Vemork had told us that no one could cross the gorge. But we doubted that. Local people often do not know their own area all that well, when you boil it right down. On the last day before the attack, I left the others to investigate on my own the possibility of getting across the river gorge. In the bright daylight I strolled down the country road past Vaaer.[4]

Haukelid picks up the narrative: 'Thursday 25th February, we decided to move forward during Saturday to a jumping off position as near Vemork as possible. We were to strike on Sunday night. As an advance base Jens [Poulsson] recommended the Fjosbudal, where there were isolated summer farms and huts which were seldom or never used. We were to spend Friday night in a hut by Lake Langesjaa, where Jens knew there would be plenty of good food. It belonged to an ironmonger in Rjukan, but even if he had fitted it with iron doors and shutters, we should get in all right with our explosives!'[5] These wouldn't in fact be necessary, Ronneberg had taken the wise precaution of bringing a pair of heavy duty bolt croppers in his pack. These could apparently 'go through three quarters of an inch of iron like butter'. And they did. Kjelstrup became deeply infatuated with these top end cutters, his own standard issue had saved his life by deflecting an enemy round in 1940, so this was quite understandable.

The Commandos made free with the quite generous supplies left in the shieling, including commandeering an unexpected but much appreciated unopened bottle of Scotch. Poulsson, the local lad, was careful to limit their pillaging to only that which was strictly necessary, they weren't brigands. Kjelstrup led the team forward on the Saturday morning as they filed, nine white clad wraiths, through the Jaerbudal. To be detected would be fatal and many from Rjukan had huts up there. But they passed through apparently unseen, as though war had not yet reached

Austrian mountain troops deploy across Norway in 1940. (Archives of Nordland)

German forces come face to face with a blown bridge in Nordland, typical of the sabotage efforts they would face during the battles for Norway. (Archives of Nordland)

Vidkun Quisling, nominal head of state within occupied Norway, inspects the troops in Skien, Telemark, in 1942. (Arkivverket)

Quisling meets with Adolf Hitler on a state visit to Berlin in 1942. (Arkivverket)

Norwegian Quisling commanders visit Norwegian SS in 1943. (Arkivverket)

Josef Terboven (front), *Reichkommissar* for occupied Norway, delivers a speech. Behind him, left to right, are *Kriegsmarine Generaladmiral* Hermann Boehm, propaganda minister Georg Wilhelm Müller, invasion architect Nikolaus von Falkenhorst, Luftwaffe General Willi Harmjanz, SS-*Obergruppenführer* Wilhelm Rediess and *Generalleutnant* Hans von Kempski. (Riksarkivet)

The Grini concentration camp, the largest in Norway throughout the occupation. Five men of Operation *Freshman* were executed there on 18 January 1943. (Library of Congress)

Queues for food in Oslo, 1942. Shortages caused suffering among the general populace. (Library of Congress)

Norwegian refugees make their way across the porous border with Sweden, escaping the occupation. (Arkivverket)

Young Norwegian men in Oreyd refugee camp, Sweden. (Arkivverket)

Above: The imposing Vemork power plant outside Rjukan, Telemark. Vemork was the first plant in the world to mass-produce heavy water. (National Library of Norway)

Below: Svelgfoss power plant in Notodden, Telemark. (Norwegian Folk Museum)

Above: Saheim power plant (or Rjukan II) in Rjukan, Telemark. (Norwegian Folk Museum)

Below: Pipelines at Rjukan demonstrate the scale of the operation. (Norwegian Folk Museum)

Leif Tronstad in 1940, probably photographed after the invasion. By this time Tronstad had already organised a line of defence with volunteer marksmen covering the retreat of Norwegian forces at Østerdalen, and he was rapidly involved in resistance efforts. (National Library of Norway)

Kompani Linge in 1945. Back, left to right: Oddvar Dobloug, Erling Jensen and Joachim Rønneberg. Front, left to right: Martin Olsen, Birger Rasmussen, Ragnar Ulstein and William Johannesen. (Musea i Nord-Østerdalen)

Tronstad (bottom left) along with, left to right, King Haakon VII, Harald Sandvik and Crown Prince Olav at Glenmore Lodge, Scotland, where Kompani Linge were trained. (Henrik Aamodt)

Opposite: The site of the Fieldfare cabin used by the saboteurs. The original cabin can be seen in the picture upon patient examination. (Frankemann)

Above: SF *Hydro*, the steam-powered railway ferry used to transport materials from Rjukan to the port at Skien. The saboteurs sank it in the deepest part of Lake Tinn, dealing a massive blow to Nazi efforts to stockpile heavy water. (Norway National Library)

Below: DF *Galtesund*, which saboteur Einar Skinnarland hijacked in 1942 in order to reach Britain, whereupon he was immediately enrolled in Kompani Linge. (Norway National Library)

Above and below: Aftermath. Fascist collaborator Vidkun Quisling in custody (above) and about to meet his end (below) before the firing squad in 1945. (Arkivverket/National Library of Norway)

Above: From left to right, Knut Haukelid, Joachim Ronneberg, Jens-Anton Poulsson and Kasper Idland greet King Haakon VII in 1948 at the premiere of *Operation Swallow*, a film about the heavy water sabotage. (Oslo Museum)

Right: Claus Helberg in later life. (Glomdal Museum)

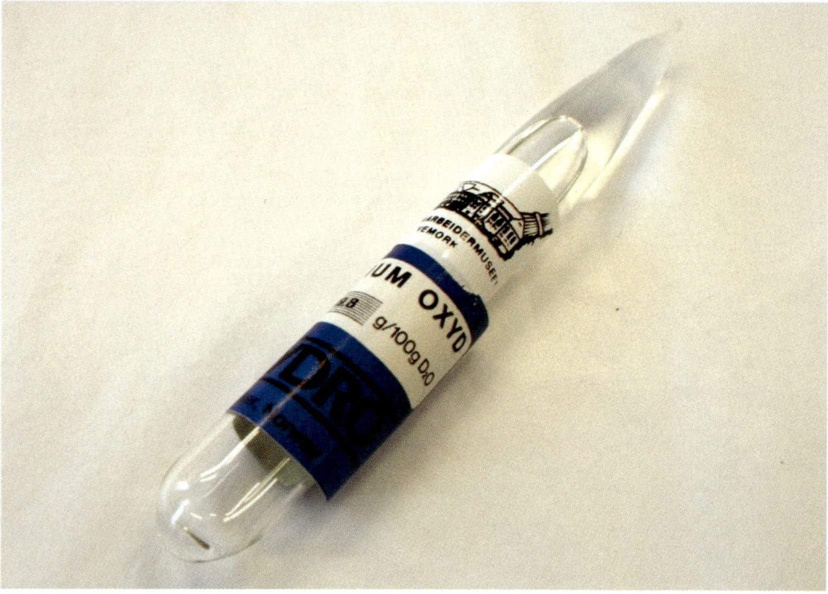

Above: An ampoule of heavy water from the Second World War era. (Aust-Agder Museum and Archive)

Below: A modern photograph of Nore power plant, another targeted during the Second World War. (Peltonman)

Opposite: Vemork power plant today. It has been converted into an industrial museum which explores the sabotage operations along with a history of the labour movement. Another power plant, opened in 1971, is built inside the mountain behind the old one. (Efarestv)

A statue depicting Joachim Ronneberg and commemorating the work of the saboteurs, in Alesund, Norway. (Olaf Meister)

this frozen and remote upland. They were, of course, about to change all of that.

Helberg was still off scouting the town, making use of a secret source he had there – clearly the less the rest of *Gunnerside* knew about this contact, the safer they'd be. Some, including most vociferously Knut Haugland, thought the ravine impracticable, as the Germans themselves had concluded, yet aerial photographs suggested it might just be doable. It would be down to the redoubtable Helberg to find out. Off he went back towards Rjukan. If anyone could find a viable route it had to be the arch swashbuckler:

> A little farther down, I left the road and headed for the gorge. I was able to make it down to the river ice, and I was also lucky enough to find a somewhat passable way up the factory side. It would not be as easy to go this way in the dark, but it was better than trying to get past the German guards on the bridge. On the basis of my information on the way across the gorge, our plan for our advance and attack was simple. As it got dark, we would start down the Fjosbu Valley, and then hike down into the Vestfjord Valley until we reached the country road. We would then follow the road about one kilometre to the spot where I had found the ford, move across the river and up on to the railroad line between Rjukan and Vemork on the south side of the valley.[6]

As impressive at Von Falkenhorst's defences for his precious gems might be, if the Commandos could scale the sheer cliffs, then the single weakest link was exposed – the seldom used railway line hacked out of the rock that fed the plant. It was unguarded:

> We would follow the railroad up to the fence surrounding the factory area. After cutting our way through, we would split into a cover party and a blasting party. The cover party would take up positions so it could make a surprise attack in case the Germans sounded the alarm. If all stayed quiet, everyone was to remain at their posts until the explosion was heard. If

> the alarm was sounded during the advance, the cover party would attack the guards immediately. When the explosion was heard, they should assume that the blasting party was outside the factory area.[7]
>
> The Germans had a guard of 30 men stationed in a hut in the middle of the yard between the buildings. We reckoned that from 8 to 12 of this force would be in the hut, the others being distributed around the area of the works. One double guard on the bridge over the ravine, another in the guardroom at the entrance. One or two men would be upon the height above Vaaer to guard the reservoir which supplied the power station.[8]
>
> Our plan as usual on such occasions, was to divide our force in two – a blowing up party and a covering party. The leader of the operation, Joachim [Ronneberg] was to be in command of the blowing up party, consisting of Frederik, Birger and Kasper. The covering party with Jens, Arne, Claus and the Chicken was to be under my command. The blowing-up party would confine itself absolutely to the carrying out of the task; while the covering party was to deal with any enemy resistance, by force of arms if necessary.[9]

Haukelid in his memoir even finds time to chide von Falkenhorst for not beefing up the guard, given the huge number of soldiers under his overall command! The blow about to be struck would make a mockery of them all.

'An order to withdraw would then be given. The passwords would be: 'Piccadilly? Leicester Square!' The blasting party would destroy the heavy water plant in the basement of the Hydrogen Factory. They would try to find an open door into the factory. If they did not succeed, they would try to crawl through a tunnel for pipes and cables leading into the building. If something unforeseen happened, everyone was to act on his own initiative to complete the mission ... It was overcast and there were heavy winds when we went down into the valley with skis and packs.'[10]

In the final run up to the attack, the saboteurs pondered the best route out, as well as in. This was no suicide mission, it was a surgical strike, delivered with crushing impact yet allowing the raiders to slip back into the night and avoid the massive net that was bound to fall. Tronstad's original brief had been for them to stash packs and heavy gear at Vaaer, then tab down off the high ground with only personal weapons and explosive devices. They'd bash a track down through the heavy snow and leave this as an avenue for their getaway. A second alternative which had now presented itself was to descend via the Fjosbudal, but exfiltrating this way meant they'd have to ford or cross the river and the bridge at Vaaer. The obvious weakness of this was that having carried out the demolitions, the German defenders would be thoroughly alerted and push up from Rjukan to seal off all exits (see map three).

Now, it was feasible to withdraw over the suspension bridge, though that certainly meant having to deal with the guards, or they could slither back down the way they'd come and proceed along the Ryes Road. This would give access to a largely overlooked minor trackway that provided a service link to the mountain railway, having firstly been laid to facilitate its construction. This zig zag route, though likely to be fearsomely hard work, was preferable to trying to crash straight up the mountainside through a heavy fall. The team would struggle to maintain cohesion. As an additional factor, they worked out that the Ryes Road could be reached via an unused lane that ascended the heights on the northern flank, parallel to the main highway – *alea iacta est*...

Nobody was in any doubt as to their inevitable fate if captured. The *Freshman* disaster had clearly signalled the full import of Hitler's command order. If wounded, you'd be left with the cold winter's wind and your suicide pill. 'We were having a smoke while waiting,' Haukelid recalled, 'when we noticed that some young people from Rjukan had had the idea of making a trip to the Fjosbudal. The next hut, which was not many yards from ours, had been occupied by two couples. They were taken into custody and told to stay in the hut till Sunday evening.'[11] These youngsters were clearly very surprised to be accosted by fellow Norwegians in uniform. The war was not so far away after all!

At 20.00 hours, they set off down the steep slope. About halfway, they sighted the plant itself, the demonic grail of their quest, 'like a medieval castle' built in a wholly inaccessible landscape. Haukelid could now see why the Germans only felt a modest garrison was required. The hum and whirr of the factory's gear came fitfully on the wind.

> We all wore British uniforms. On the road, the ice was as slick and shiny as an ice rink. Off and on, the wind came in such strong gusts that we fell on the road. The weather could not have been better. We were the only ones out in such conditions.[12]
>
> When we approached the high road between Rjukan and the Mosvatn [dam] we found it best to carry our skis. We joined the road a little above the bend at Vaaer. Claus [Helberg] had reported that the Germans for some reason or other had moved a fairly large number of troops towards the Mosvatn in the course of the day. We were therefore rather nervous when we came down onto the road, but we had to keep to it if we were to arrive in anything like reasonable time. The hillsides were so steep, and so much snow was lying, that it was impossible for us to get forward under cover of the wood. So, we advanced in single file, with Claus well ahead as scout.[13]

Darkness was their friend. The Germans might be prepared – but not *that* prepared. They marched along the high road for just over a mile before they reached the bends at Vaaer. These they simply cut out by moving cross country to come out again by the Vaaer Beck, this also avoided having to sneak through the village itself. Descending was like sliding through a chute lined with slush, so wet and weighty the snow.

About half the party had come up to the Vaaer bridge when two service buses appeared, their lights cutting through the dark, shift workers on their way to the plant. Some of the Commandos could easily have slithered helplessly off the wet slop to thud onto the moving roof of one of the vehicles, very *Where Eagles Dare* but potentially disastrous. Happily, nobody did and the workers, lost in the humdrum of just another nightshift, trundled oblivious on their way. We wonder many of those, reading post-war accounts, realised that was their shift!

By now it was nearly midnight and having regrouped in the lee of a small shack on the east bank of the river, they set of again in Indian file. Despite wartime restrictions, local traffic tended to move with all lights blazing, a blessing as they could detect any vehicle approaching with ample warning. They kept going steadily for another couple of miles, Helberg still pathfinder, until by a small enclosure he motioned a halt. This was the crossing place where they'd get over the Maan River and begin their epic climb. 'We ran across the ice and began to clamber up the other side. In such circumstances it is always tiring to be one of a party. One has a feeling all the time that one's comrades are climbing faster than oneself, and that one must hurry to keep up with them. This naturally means the others quicken their pace too, and the speed of the whole party steadily increases. It was a wearisome climb before we stood on the railway line above.'[14]

Nonetheless, Helberg was fully vindicated in his tactical assessment:

> The climb down into the gorge went better than expected, but the river had risen because of the mild weather, and the water ran high above the river ice. But we got across and up to the railroad on the other side. When we were only 500 metres from our target, we stopped to go through the plan for the last time. It was vital that each man knew his assignment. We had a fine view of the bridge where the guards were posted. There would be a changing of guards at midnight, and because the guards' house was up by the factory building, we let the off-duty guards have 30 minutes to fall asleep.[15]

Nobody on the enemy side had noticed their totally unorthodox approach and breathtaking ascent. The line, only lightly dusted in snow, led straight and clear. There was no moon. Any noise they created was drowned out by the brisk south-westerly. A little before the witching hour they came across a railwaymen's shed. So far so good. Here they paused briefly to eat part of their chocolate ration and held back, waiting for the sentries to change shifts. Once they stopped, the wind became more cutting. At midnight, the Germans, ever punctual, would swap over and half an hour later, they'd go in, covering party leading to cut their way, if this proved necessary, into the compound.

Haukelid, 'went forward with Arne [Kjelstrup] a few yards behind me ... We had calculated that the Germans would have laid minefields to protect the factory. If we were unlucky enough to come upon a mine, then not more than two of us would go in the first round. Arne and I found a trodden path which followed the railway line from a tool shed right up to the factory. Here we felt fairly safe.' Kjelstrup had the cutters and quickly disposed of the thin chain. They were into 'one of the most important military objectives in Europe'.[16]

Haukelid inventoried their weapons, all locked, loaded and cocked for the assault, five Thompsons, two Lee Enfield No. IV sniper rifles, plus handguns, fighting knives and grenades. 'Then we went up to the fence and cut our way through. I was with the cover party, and we took up our positions. We covered the German guardhouse and the entrances of the factory. Fortunately, we did not see a single guard. The noise from the factory was quite loud, and there was no other sound.'

Ronneberg led the demolition group towards the basement doors, but all were thoroughly secured. This meant the cable duct, and while he and Kayser groped for this they got separated from the other two:

> It was pitch dark, and, even if a German had come, I do not believe he would have seen us leaning against the factory wall. The blasting party had some difficulties getting into the factory. The doors were locked, and it was not easy to find the cable tunnel in the dark. During the search, the men lost track of each other, and only two finally found the opening. They had the explosives, and the tunnel was just wide enough for them to crawl through.

> They entered the room next to the heavy water plant. The entrance door was open, and they walked right in. There was a worker there, and he probably was scared stiff at the sight of the uniformed men. But he sensed the situation, and he kept quiet ... The men started to place their explosives. They had practised on wooden models in England and loads and fuses had been cut to size beforehand. While they were busy working, the silence was broken by the sound of glass being shattered nearby. One of the others in the blasting party who had not found the cable opening had decided to act on his own, and, when he found a window, he smashed it and crawled through.

In fact there were two, Stromsheim and Idland.[17]

Kayser spoke to the very nervous watchman, pointing out their British uniforms and kit: 'I don't really expect there are many of the master race in Rjukan who have had the chance of getting so close to an Allied soldier!'[18] This was important, ensuring that the enemy were completely convinced these raiders were British and not Norwegian. Any suspicion of local involvement and vicious reprisals were sure to follow.

Ronneberg cut the fuses down to 30 seconds. The hapless technician suddenly blurted out he'd lost his spectacles (almost impossible to replace in wartime conditions). 'On your nose,' Ronneberg pointed out. The watchman was released through the internal door and advised to run as fast as he could. He didn't need any further urging.

> He [Stromsheim] assisted in completing the work. To make sure that the explosives would not be removed, the fuses were cut as short as possible. The leader lighted the fuse, the worker dashed up the stairs to the upper floors, and the saboteurs ran out through one of the basement doors. The loads exploded 30 seconds later, and the heavy water instruments were destroyed. Except for smashed windows, only the instruments were demolished. The factory itself was undamaged. It may have been this fact that made the Germans react so slowly.

> One of the off-duty guards poked his head out the guardhouse door, turned a flashlight on the factory building, and then went back to bed! Thus, we got an unexpectedly large head start, and it was not until we were up on the country road that the first cars with Germans came up from Rjukan. We found our skis and headed for the town.[19]

Nerves amongst the covering party were getting frayed, hardly surprising. They were in the centre of a well defended and vitally important enemy controlled facility; seconds passed like hours. How long could their luck last? 'The time seemed long to us who stood waiting outside. The Chicken had followed the blowing-up party as far as he could and was to see that none of the sentries at the gate or the bridge could come up towards the factory area. Arne stood in readiness to deny access to the sentries on the height above Vaaer. Claus had the task of guarding the road at the point where we had forced our way into the factory area and covering our retreat.'[20]

Poulsson and Haukelid were covering the guard hut, the electric light plant was on their right, the guardroom straight ahead. Any sign of reaction and they'd close in on the place, Thompsons blazing. Haukelid was also carrying a stash of chloroform pads to neutralise any Norwegian workmen unlucky enough to pitch up. There were none and the pair crouched about 20 metres from the hut, concealed behind some handily placed drums. The hum, whirr and clatter of electrical gear from the plant was a blessing. Poulsson and Haukelid could converse normally with no fear of their being overheard. Dwarfed by the hulking shadow of these industrial colossi, they could only wait. They'd no idea how the demolition party were faring but, well equipped with automatic weapons and grenades, were more than ready if the guards became restive: 'You must remember to call out Heil Hitler when you open the door and throw the bombs!'

Both men knew they could rain down death and destruction on the as yet unsuspecting Germans inside their flimsy shed. Haukelid fondly remembered a very similar skirmish during an earlier fight when the Norwegians had surrounded an enemy section holding out in a timber house, 'dead Germans hanging out of the window, and dead Germans lying inside before we had finished shooting the house to pieces.'[21] After 20 minutes or so they heard a distinct but muffled bang. Is that it? It seemed a bit of an anti-climax, they'd expected something more dramatic.

Luckily for the saboteurs, the odd smash and bang wasn't unusual at Vemork, and no great Teutonic horde came screaming out into the night bent on retribution. Nothing happened. After a few minutes a lone, not very attentive sentry wandered out. He checked the external doors were locked. He seemed to be both unarmed and uninterested, and soon went back inside, but he emerged more purposefully a moment later, this time bearing a torch. Haukelid awaited Poulsson's order to open up but after flashing the beam around quite cursorily, the sentry returned to his warm stove. And lived to tell his grandchildren.

All of the attackers and covering party were withdrawing back to their RV at the breached gate. Kjelstrup took the use of their agreed password 'Piccadilly' seriously. But Poulsson and Haukelid ignored the correct response, 'Leicester Square'. Plans, however well laid and intentioned, seldom survive contact with reality. Kjelstrup was offended but was simply advised to 'shut up!' Helberg was still on point but found the demolitions group had already reached the railway.

Poulsson suggested Haukelid dump the chloroform as it would obviously not be needed. He had already done so, fearful an ampoule might smash and he'd knock himself out! He'd been careful to wipe the tubes as the enemy already had his prints on file from 1940. Poulsson pretended mock horror, as he'd unleashed gas warfare! Pretty tame considering what they'd just destroyed.

According to Helberg, 'There was some discussion as to whether we should go straight up the mountain slope, but those of us who knew the area were able to veto that. The slope would have been too steep and heavy. Of course, it was risky to go down this road, because we could run into Germans at any moment, but we had no other way. The German cars went towards Vemork on the road right below us. Nobody must have thought that we would go toward Rjukan. At 0500, we were at Gvepseborg. And we felt great satisfaction when we sat and looked at Vemork and thought of all the commotion we had caused down there.'[22]

Only when they'd scrambled and slithered down the cliff face into the gorge did they hear the first wail of sirens. Cars skittered over the road, panic in their very motion. Torches flickered erratically along the line of the railway above. Their means of entry and escape had been detected but the foxes had long since fled the hen coop. The hounds were blundering. Now it was a test of stamina; already worn out by their earlier exertions, adrenalin kept them going, exhilaration crept through their veins. Helberg with Poulsson, local boys both, led the trail. Off by a sandpit just before the cable car station, they climbed through trees to reach the Ryes Road. No probing searchlight fingers tracked them.

Still, it took them another three hours to get up the road, their loads seeming to increase in weight. At last, as they ascended the lip of the Hardanger, the plateau greeted them in its usual raucous way with a howling head wind. Terrible conditions, but the tempest was on their side, tracks were obliterated, pursuit near impossible. More tough going awaited them tilluntil much later in the new day they struggled into the hut at Lake Langesjaa.

Mission accomplished.

8

ESCAPE!

'What's necessary will always be accomplished.'

Heroes of Telemark

HIGH CONCENTRATION INSTALLATION AT VEMORK COMPLETELY DESTROYED ON NIGHT OF 27TH–28TH STOP *GUNNERSIDE* HAS GONE TO SWEDEN STOP GREETINGS STOP

(Sent that evening.)

The Hardanger's inhospitable frenzy was now their ally. Whilst it made their return to the Svensbu shed yet another marathon, it blotted out any hopes the Germans might have had of mounting an effective hot trod. Von Falkenhorst's crown jewels had been vapourised under their very noses and heads were sure to roll. Alf Larsen, chief engineer, was one of the first to assess the totality of destruction: each of the high-concentration cells had been effectively sabotaged, wreckage sprayed by water spurting from pipes burst by shrapnel. Whereas damage was contained, it was nonetheless total.

General Wilhelm Rediess arrived on Sunday. He was not a happy man. Fallout from the disaster would inevitably pass up the command chain and he was keen to ensure somebody else carried the can. He did appear convinced by the carefully laid trail of evidence, pointing squarely at a British-led raid. Even if the watchman twigged these raiders were in fact Norwegian, he had the good sense to keep that knowledge to himself. Rediess could see how the Commandos had got into the yard and their footprints were still clear, blood had also leaked from an injury to his hand Ronneberg had sustained. He realised the raid could not have been carried out so effectively without detailed insider knowledge – Brun was the obvious suspect, but he was long gone. In his fury, Rediess resorted to brute force and terror, his standard toolkit. Only von Falkenhorst's arrival prevented reprisals against the plant's civilian staff.

Despite his obvious dismay, the General was impressed: 'The most splendid coup I have seen in this war,' he grudgingly admitted. Sporting perhaps, but not what his superiors in Berlin would wish to hear. He declared that the attack was a British military operation and thus released those Norwegians Rediess had rounded up. A collective sigh of relief locally. Not only had *Gunnerside* achieved its objectives without loss and got clear away, but they had succeeded in deflecting any suspicion of complicity away from the civilian population. Now it was time to launch a massive manhunt, (*razzia*), scouring that twisted landscape of the Hardanger. This was an environment utterly alien to those German rank and file tasked with the job, who, without training in mountain warfare, would be likely to blunder about indefinitely. As they did.

As a postscript at the plant, Falkenhorst's blood pressure shot up again when the guards, attempting to initiate the floodlights, couldn't even find the switch. Instant postings to the Eastern Front. Rediess sent a rather bland report back to Germany: 'On the night of 27th–28th February 1943, about 1:15 a.m., an installation of importance to the war economy was destroyed at the Vemork factory near Rjukan by the detonation of explosive charges. The attack was carried out by three armed men wearing green-grey uniforms. They gained entrance to the factory by cutting a chain in the gate and passed both German sentries and Norwegian watchmen undetected. From the effects they left behind, it can be assumed that they came from Britain ... The sabotage was a combined British Intelligence and Norwegian Resistance operation.

Security police investigations are still proceeding into the matter.'[1] The General was all too acutely aware that trains to the Eastern Front ran all the time, and rank was not always a get out of train card.

> We headed for the Hardanger Plateau again and for new assignments. Five men made straight for Sweden on skis, in English uniforms, with guns and ammunition. Our leader made his way alone to Oslo and then on to Sweden and England. Two went westward and established headquarters for the resistance forces in the Vinje region; one was back in England by Christmas, while the other stayed in Norway until the end of the war. The two telegraph operators established a radio station in the Hamre Mountains by the Mos Lake, and one operated this station until the war was over. Many exciting stories can be told about each one of the men from that time on, but I [Helberg] will limit myself to my own adventures after our action.[2]

Claus Helberg left the rest of the group to travel back up to Fjosbudalen and we'll pick up the trail of his adventures. Now, eight men were left to escape. This meant Sweden and getting there was a major challenge in itself. With every German in the occupied land looking for them, the getaway looked set to be every bit as perilous as the mission. On 2 March, they skied off from Svensbu heading for their original landing zone at Skrykken, but hostile winds blew them back. Next day they tried again and with better luck. They had time to divert via Sletterdal where they stashed a note for their two wireless operators Skinnarland and Haughland, confirming their success. The two W/Os would remain in Norway, Skinnarland taking over the comms role while Haughland applied himself to recruitment and training.

Of the *Gunnerside* five, Ronneberg, Idland, Kayser, Stromsheim and Storhaug would attempt the 250-mile (400-kilometres) dash to the Swedish border, which Ronneberg estimated would take around ten days. They'd rely on their skis and remain in uniform, ready to shoot their way past any opposition. Once they had crossed the frontier, they'd dump all kit and weapons then, in two groups, and announce themselves as political refugees. Generally, the Swedes took a pretty relaxed approach.

They'd know just fine who these asylum seekers were but, if not confronted by obvious military symbols, wouldn't worry overmuch. Meanwhile, as noted, Haukelid and Kjelstrup would stay put on the Hardanger to conjure new mayhem while Poulsson with Helberg would travel to Oslo, deciding on future plans when they got there. That was if Helberg, off on his own temporary quest, actually turned up.

As he himself explains:

> First, I had to make a trip to Fjosbu Valley, and then I was to catch up with the others near Rjukan. But because of bad weather, I was delayed and did not find them. I therefore went alone to Geilo and took the train to Oslo. I left my uniform in the mountains. So far, the Germans had not started to search the Hardanger Plateau, and three weeks after the action I headed for the mountains again.[3]

Closing the stable door after the horse has bolted has a long history amongst those who, having lost what they treasured, endeavour to ensure nobody can steal it a second time, even if the prize is gone. Wanted posters showing some generically 'Aryan' types went up, much to the locals' secret mirth. Efforts began to repair and replace the sabotaged installation and that massive manhunt was initiated.

Ronneberg should indeed have shot Kristiansen; he was the loose end that could unravel all. The idiot had shared his chocolate bounty with local kids and the tell-tale wrappers had proved his undoing. Local police nicked him, and he blurted out the whole story of his captivity and near-death experiences. He did maintain the view (which he may well have believed), that the group who'd abducted him were all Brits. So, in the long run and more by accident, he did no harm. He was eventually released without sanction, very few men got to be that lucky twice in a row.

Getting safely across the Swedish border wasn't about travelling in a straight line; built-up areas, even hamlets and farms, were best avoided. This meant taking the hardest and most exposed route, over high ground, the remoter the better. Ronneberg had carefully mapped their intended route. They had serviceable maps and compasses. In theory, this all sounded workable, but the hostile ground could yet confuse them, and the enemy was fully on alert. Catching or killing these now notorious saboteurs would help assuage the wrath of High Command. Terboven, Rediess and von Falkenhorst all knew they really didn't need another debacle.

SOE had in fact issued *Gunnerside* with three sets of maps, all 1:100,000 scale and each of these held 26 localised larger-scale inserts. Ronneberg had noted they couldn't just head due east as they'd have to get round or through various population centres. To the south, a large lake would prove a very significant obstacle, so their route must lie to the north, turning gradually eastwards. This line had its own problems. They'd need to steer well clear of Lillehammer where the Wehrmacht maintained its skiing and mountain warfare schools: 'One set [of maps] we had pinned to the walls of the huts we stayed in before setting out and, having drawn the line of our intended route, each member of the withdrawal party was encouraged to study them so he had a good idea in his mind where we would be going.'[4]

Haughland and Skinnarland were being left behind on the Hardanger as the net began to close, this massive *razzia* the Germans mounted to sweep the area. Nonetheless, the SOE men were not overly impressed and never felt that threatened. As a precaution, they vacated their cosy hut and roughed it under canvas amongst the peaks looming over Mostvatn Dam. This was Einar's stomping ground and they watched search parties below and aircraft above, all trying to winkle them out. The resourceful pair burrowed into the snow to dig out a two-roomed ice bunker. Despite the number of men and resources deployed, the searchers scarcely brushed the uncompromising surface of the mountains. The Hardanger was just not their environment, and they probably couldn't work out how anyone could possibly survive up there.

Quisling auxiliaries were brought in from the Kongsvinger Battalion This unit was an element (3rd Battalion) of 27th SS Police Regiment, formed during 1941 and which, early in 1943, had received its SS designation. The battalion was raised from the town of Kongsvinger, a municipality within Innlandet County. Various other pro-Nazi police groups were also seconded, it didn't boost efficiency. They searched every hut (pausing only to loot whatever they could), adding to a confusion of trails rather than tracking any. Haukelid, looking contemptuously down on them through binoculars, was astonished at their useless milling. To him, the archetypal frontiersman, you had to track your prey moving like your prey. This great *razzia* was largely self-defeating; how satisfying.

In mid-April, SOE received a report from one of their agents in Oslo, with the rather unflattering codename, 'Parrot', who reported on the ongoing sweeps of the Hardanger and that quite a number of German troops had been wounded in what would now be called 'blue on blue', friendly fire incidents. Units were obviously so jumpy, they were firing at shadows, provoking firefights between their own patrols. Cack-handed as the Axis response might seem, local Norwegians bore the brunt of collateral damage. There was martial law, no free travel, German patrols looted and torched steadings at will. Wartime shortages, already biting, were being significantly exacerbated. Germans commandeered anything they needed, or fancied, while locals went without, no milk for schools and hunger- or illness-related deaths in Rjukan began to climb.

Ronneberg and the other four set off in decent conditions but loaded with heavy packs, dragging a laden sled. Weather was changeable, hard snow was easy going, wet snow far more difficult work and clear, boulder-strewn scrublands – just awful. First night out, they camped in an abandoned upland hostel where some precious foodstuffs helped eke out their meagre supplies. Next day, good weather holding, they set off on another leg. It wasn't just about progress; it was cautious progress. Two members of the group always went on ahead to scout the landscape with glasses. Detection was fatal. They needed the toboggan, but an afternoon struggle to manoeuvre the sled down a steep incline meant an equally gruelling ascent. That evening, they occupied another empty lodge by Lake Rodungen. Ronneberg's hand was sorely inflamed, and the dazzling horizon of endless peaks and troughs defied their well-honed map-reading skills. They had to navigate by compass alone.

Joachim recalled the start of their epic trek:

> We had appalling weather conditions and could not make good time ... We took only what food supplies we thought we'd need, but we were still soldiers in uniform, in accordance with the strict orders laid down in London, and our guns, ammunition and other military equipment added to the weight we had to carry. It was absolutely necessary weight though, for we knew we might encounter German soldiers at any moment.[5]

Sunday was the day of the week when many Norwegians skied for recreation. Best to lie low. There was a very nervous moment when one skier seemed to be headed straight for their hideout, happily his destination lay next door. Monday now and they stayed out of sight till light thickened for a rapid dash across the exposed breadth of the Hallingdalen Valley. So far so good, but so many paths spread out before them. Only one was right, so that night was spent in the open, huddled in damp sleeping bags.

Dawn brought high, freezing winds cutting through soggy kit. Having identified their location, the team set off at a quick march to reach another shieling a few miles distant. After drying out as best they could, the team pressed on and by dusk had got to a surprisingly well stocked steading at Fjellstolen. Next destination was eastwards towards Aurdalsfjord, but their careful precautions failed to detect a pair of skiers heading in the opposite direction. Both groups waved unconcernedly. Team *Gunnerside* were clad in artic whites with their weapons barely concealed. The other group appear, happily, to have mistaken them for a German patrol and detected nothing untoward.

Now for the Aurdalsfjord. It should have been frozen, but so much ice had melted. One point looked like it was still a viable crossing, so Ronneberg took it on himself to make the attempt. Gingerly, spreading his weight, ice axe stretched out in front to probe with, he inched forwards. He made it safely, to a collective sigh of relief. It was after midnight before they'd all got across and scaled the heights opposite. Their well-earned repose was taken in a shed owned by a man called Christiansen who'd plastered the walls with photos of the King – a patriot's eyrie. Somehow, that felt really good: a tangible reminder in this lonely, blasted spot of what they were fighting for and why they struggled so hard.

So far so good didn't last that far. Wet snow was among the worst possible conditions and as they set off the following morning, temperatures rose and the march slowed, slush clung to their skis like Somme mud. The good news was that their kit overall performed well, woollen battledress, as anyone who's worn it knows, is heavy and hot but also immensely durable. Add in a woollen shirt (the devil's invention) and, after any kind of movement, you're in your personal sauna. Your underwear yes, is also of wool manufacture, itchy and sweaty during the march but comfortably warming at night. Their rabbit skin jockstraps were not a fetish item, but they did chafe when perspiring and were prone to overheating.

Kasper Idland had always known he was by far the weakest skier and had blagged his way onto the mission hoping nobody would ask. Nobody had, but now he really struggled to keep up, yet he managed: 'Quiet Kasper Idland was extraordinarily willing and good-humoured during our retreat ... He tried very hard to keep up with us. He kept struggling to keep up, but never complained, never asked us to slow down. If we had left him behind, he would not have called out. We knew that, but of course we had no intention of leaving him behind.'[6] (In the words of the fictional Rear-Admiral George Tarrant – and Ronald Reagan – where do we find such men?)

These were men of steel, but they still needed sustenance, a lot of it given their huge exertions, and supplies were running out. Their mantra was to solemnly burn each of their 26 local maps once they were clear of the locale. Deliverance got closer and closer by the day. Axis planes passed overhead but showed no interest in them, they'd passed well beyond the net of the *razzia*. No shielings lay across this current line of their escape, so they were forced to spend shivering nights huddled in the open. Still, with food shortages now critical, they had to seek out and plunder such steadings as there were. The haul was, every time, meagre. Sweden grew daily nearer but the border area was alive with German and Quisling patrols, maximum caution was needed. To be caught at this juncture with salvation so close would be agony.

Next obstacle was the wide River Glomma. They proposed to cross close, but not too close, to the town of Rena. Storhaug, a local born in the settlement, volunteered to be the guide as he knew the area well. Not as well as he thought: they ended up four miles (six kilometres) off course. Storhaug wasn't popular but Ronneberg, when he'd cooled down, conceded that orientation in this maze of narrow valleys was tricky, at least on the mountain you got an overview. They did finally find the river but dismayingly, it was free of ice.

Storhaug redeemed himself when he roused a friend he knew to be staunch. The man was indeed a patriot and, moreover, one who had a dinghy. Safely over the water they were able to hide in woodland. Forested hills kept them pretty much invisible for the next couple of days, though they had to sleep out in the open. Cold and hunger were their attendant demons, gnawing away at their reserves of stamina, already badly overdrawn. For all the remaining distance, the ground proved especially hard, wild rock and stunted trees. But at a quarter past ten on the evening of 18 March, they successfully crossed the finish line. After 15 days hellish marching, they'd done it. *Gunnerside* was coming home. They all shook hands. That's so Norwegian, it's almost British.

Now they might be out of the fjords, but they weren't clear of the woods. Still cold, still very hungry and with another night in the open, except this time they could light a fire, a beacon of their imminent salvation. All their military uniform, weapons and kit that had served so faithfully had to be discarded and buried. They shrugged into the civvies stashed in their packs. They were no longer warriors but refugees, even if they still looked like soldiers, albeit unshaven, unkempt and unfragrant. They kept their packs but cleared out anything that could identify them as British servicemen.

Food. They needed food urgently. Exhausted and malnourished, they were almost at the point of collapse. They also needed a good cover story. Using their training base in Scotland as a template, they'd describe how they had been conscripted by the Nazis as forced labour to build a new set of barracks and ancillary buildings. They had never met before, but they formed a resistance cell within the site. Fearful of discovery, they'd staged a breakout and made a run for the frontier.

Their first day in neutral Sweden proved relatively relaxed. They had marched east for 12 miles (20 kilometres) before a border patrol picked them up. They were deposited with the local sheriff, who proved friendly and had them hospitalised for a general service and makeover. They weren't under arrest as such. Bathed, with bliss-inducing shaves and haircuts, at ease in clean clothes, they spent that evening dining in a local restaurant, no blackout, no rationing and no Nazis. It was time to celebrate.

Their cover story easily stood the test of a very superficial interrogation, and they were free to travel within Sweden without restraint. The authorities probably knew who they were or certainly guessed. After travelling first to Kjester, they reached Stockholm and the British Legation without further incident. This must have seemed quite banal after the tribulations they had endured to get there. SOE in the Swedish capital, rather wary of their visitors at first, radioed the glad tidings back to Baker Street. March melted into April, and they finally took ship from Bromma back into the UK and then by train to London.

As they were in Britain, the first thing they were offered on landing was a nice cup of tea. Ronneberg decided it was good to be 'home': 'Before leaving Sweden we bought all new skiing and camping equipment, compasses, sleeping bags, extra woollens and so forth, in preparation for our next assignment from Britain to Norway. The biggest thrill of all, though, for me, was our return to Britain, which I considered my second home. In fact, during the operation, we always liked to talk about going home – to Britain. And yet, while in Britain, we always talked about going home – to Norway.'[7]

When Poulsson set off to reach Oslo, there was still been no sign of Helberg, and he feared for his friend's survival, but was bolstered by the knowledge Helberg had a genius both for landing himself in hot water and for springing clear of it. It was an odd sensation for Poulsson, having bidden farewell to the escape party, shaken hands with Haukelid and Kjelstrup as they pushed off on their own fresh odyssey, as he dumped his uniform and all military gear bar a pistol (origin and supply untraceable). In civvies now for the first time in a very long time, he felt like an imposter, almost an alien amongst his own folk.

Having got as far as the Uvdal valley, he took the seemingly reckless chance of booking in to a local inn. It had been some time since he had sat in front of a warm fire or ate a cooked dinner. Later, he wallowed in the luxury of a hot bath, his first in months. It could have been his last.

Dragged from a soothing revery, he became aware of voices downstairs, a random police search? Quick thinking. No time to flee, his kit, most of which he'd been washing, was strewn about. Concealing his weapon, he opened up in response to repeated knocking.

In came the local sheriff and his hulking deputy, bound to be Quislings, surely? They demanded ID so he handed over his first-class fake papers and passport. Neither of the two lawmen looked unduly stressed, if they'd been aware of their inexorable fate had they proved overly suspicious, they most certainly would have. They told him about the raid at Vemork, thus the heightened security. He stayed politely noncommittal. They looked at his bespoke sleeping bag and very un-Norwegian boots but left, apparently satisfied. Poulsson spent an uneasy night ready for fight and/or flight, but dawn came round without further incident and he was able to enjoy a hearty breakfast.

By late afternoon on 8 March, Jens Poulsson was in the heart of his nation's capital. He recorded that the Germans at this time were generally relatively relaxed, and Oslo didn't yet have the feel of a city groaning beneath the jackboot. That would come later. He stood in the pallid light of nearly spring beneath a large railway clock by the underground station entrance. Across the street was Majorstua café, his original RV with the missing Claus Helberg, from whom he'd heard nothing. Their meet had been set for the previous day, but Helberg was a no-show. Even if the Hun wasn't being unduly heavy-handed, he'd still have eyes everywhere and those eyes picked up on patterns of behaviour. That had been thoroughly drummed into them all during SOE training.

Tram wires cast their spider's web above Poulsson's head as he sauntered across the narrow cobbled lane, glancing in through the café windows. And there sat Helberg, calm and unnoticed. 'Hello Claus,' he greeted. Both men felt a rush of emotion at their reunion, the digest of shared experience and enormous dangers, but they met simply like old friends out for a coffee, or what passed for it during wartime rationing. Despite this outward serenity, their war was very far from over: they had a joint mission to Notodden where they would liaise with Milorg.

Helberg was keen to stay in Norway, less keen on Sweden – he had unfortunate form there. Anxious to retrieve the dump of weapons and gear they'd left behind on the plateau, he nonetheless had to wait a few weeks while the heat unleashed there died down. Deeming it safe and relying on intel supplied by Milorg, he left Oslo on 22 March. Poulsson was headed first for Sweden and from there the UK, where he would recuperate for a while.

Now Helberg describes his return to the uplands: 'I went from Imingen north to Skrykken. The purpose of my trip was to hide some weapons and explosives that had been left behind by *Gunnerside* at Skrykken. I found the cabin where the men had stayed and went in to get a little food after a long hike. The cabin looked like a pigsty, and it was obvious that the Germans had been there.'[8] His info was wrong, and the enemy was still swarming.

Blissfully unaware, he had a solid kip, 14 hours' worth and by 25 March had reached the Jansbu shack at Skrykken. There he experienced a nasty shock; the place had been systematically trashed. A stab of alarm.

> Perhaps they were close by! I ran out. Sure enough. Four or five men came running up towards the cabin. I put my skis on in a hurry and took off. My only weapon was a pistol. I had to rely on my abilities as a skier. After some shooting, one of the Germans was catching up with me. We were heading westward at great speed. I just wanted to have the evening sun in my face, to blind the Germans when they were shooting. But one German kept my speed. Fortunately, he also just had a pistol. Thus, the chances were equal, and the best shooter would win. The German evidently determined that the fight would be to my advantage.[9]

This was the classic, almost Western-style gun duel enacted by Richard Harris in the movie, the only time the saboteurs actually had to use their weapons.

> He became so scared when I began to shoot at him that he emptied his whole magazine at me. All misses. Suddenly, our roles were changed. He threw himself around and headed back towards the other Germans. I hesitated for a moment, and then I pursued him as fast as I could. Every second was precious. His friends could show up over the crest at any moment. At a distance of about 30 metres, I fired a few shots at him. The sun had gone down, and darkness was approaching. I was safe, at least for the time being. But the ski tracks were a problem. The Germans could follow them, even in the dark. I knew there was no snow on the ice of Vra Lake, so I made for the lake. It was a starry night, but it was dark, and I had the bad luck of skiing off a precipice in the Slette Valley, just north of Vra Lake. I hurt my left arm badly, but I had to move on.[10]

He had broken his arm, though not his skis. It was now dark, but he had to continue as best he could, only now desperately aware of how rapidly the net was tightening:

> Down by Mogen, I barely avoided the Germans. They were swarming all over the place. It was obvious that they had started a major raid. But with a broken arm, I could not head for the mountains again. I had to see a doctor. Later in the day, I reached Hamaren, tired and hungry. I knew the people there, but I could not stay because the Germans searched every house at the Mos Beach. After I had had something to eat and rested a little. I moved across Mos Lake and down to Rauland. I had been on my feet for 36 hours and covered 160 kilometres.

Bad choice; just a few miles short of his objective he ran smack into a Hun patrol. Helberg, as he'd demonstrated so many times, had the happy knack of blagging his way out of difficulty. He succeeded in convincing the Germans he was a Norwegian civilian auxiliary who had been helping them out when he sustained a fall. Happily, his fake ID and casual cool passed the test but, 'Rauland was standing on end. I could not have chosen a worse haven. There were about 300 German troops there. The weather turned mild, and it was impossible for me to consider moving on. I knew a merchant at Austbo, and I moved in with him and was allowed to sleep on his kitchen floor.'[11] He was Daniel in the lion's den but played his bluff like a master, which indeed he was. Making new friends amongst the bored Axis incomers, one even applied first aid to his damaged limb and arranged for him to hitch a ride in an ambulance to Dalen, en route to Oslo for proper medical attention.

He found himself at a local hotel, the Bandaksli, dined well and retired early. This proved one piece of bravado too far: Germans had requisitioned the other rooms.

> *Fast Thinking* During the night, I concocted a story to tell the Germans, and they bought it! I said that I had reported as a local scout for the Germans in their search for the Vemork

> saboteurs and had broken my arm in their service. The Germans must have thought that I was quite a likeable fellow. The German field doctor at Rauland examined my arm and sent me by ambulance down to Dalen, where I could travel on to Oslo by myself. In Dalen, I said *auf wiedersehen* to my helpers, and we parted as good friends. A boat was to leave the next day, and I wanted to rest a little. The tourist hotel looked inviting, and I found lodgings there. But bad luck struck again. Two German officers and their staffs came to the hotel in the evening and requisitioned most of the rooms. All the doors were blocked, and it was out of the question to try to escape.[12]

It turned out the hotel's German 'guests' included no lesser person than the strutting ogre Terboven himself, with a gaggle of senior officers hanging on. One very attractive young woman, resident in the hotel, refused to be intimidated by the *Reichskommissar*'s heavy handed gallantries and declined his imperious seduction. Terboven' s mountainous ego was offended, and he decided on spiteful retaliation.

> All 30 guests were interrogated. We each had to give our name, address, age, and purpose of staying in Dalen. My identification card, which had been made in London, withstood a close examination. Under arrest the next morning, 18 guests, including myself, were told to pack. We were arrested because of 'insolent behaviour'. I was in great doubt as to what to do, but I finally decided to go with the prisoner transport and then make a run for it. At any rate, I could not go all the way to Grini prison camp. I had put my pistol under the lining of my ski jacket. Before long, a big bus pulled up in front of the hotel, and we were accompanied to it, one by one. I made sure I was the last one to get out of the hotel, so that I would get a seat right by the exit door in the bus.[13] ... A German thought that I moved too slowly, and he gave me a kick in the rear so that I stumbled on the stairs. My pistol fell out of my ski jacket and landed between the feet of a German. Now

> I was in a jam! The Germans obviously did not know what to do with me, but the order was to send us to Grini, and an order is an order. One German guard was sitting in the front of the bus, and four others on two motorcycles convoyed us, one motorcycle 10 metres in front of and the other 10 metres behind the bus. My seat was on the floor in the rear of the bus. A young lady from Oslo sat next to me. We got to talking and actually had a cosy time. The German guard in the bus was envious, and eventually we changed seats.[14]

Lesser men might have been daunted but Helberg was always equal to the challenge: 'I got to sit up front by the exit door, and he sat in the back. Of course, that made things easy. We had driven for nine hours, and it had become dark. In the Lier Hills between Drammen and Oslo, I opened the door and jumped out, amid the oaths of the Germans. I fell on the road but got back on my feet just as the Germans in the rear motorcycle were about to jump on me. Pursued by exploding hand grenades, I got into the forest, and, thanks to the darkness, my pursuers lost me. They gave up and had to go on without me. After benevolent treatment at Lier Asylum and a three-week stay at Drammen Hospital, I went on to England through Sweden.'[15] A very close shave even by Helberg's swashbuckling standards.

After all these adventures and against all odds, *Gunnerside* was safe. Main Mission now accomplished; nearly accomplished anyway, the curtain hadn't quite fallen, not just yet. Another nuclear chasm was yawning.

9

SF *HYDRO*

Shrill as a banshee's wail of terror,
Low like an oft-heard funeral dirge,
Rising and falling without an error,
Sirens scream – Take Cover – Urge…

Ears attuned to engines droning,
Eyes upturned towards the sky,
Children sobbing, voices moaning,
Again, the death from up on high.
Explosions shake familiar buildings,
Dust and debris over all,
Bombs are falling, mangling, killing,
Voices for their loved one's call.

J.W., 'Air Raid'

We, people of my generation in this country and all who are younger have not had to experience an air raid. Terrible events being enacted in Ukraine at the time of writing, the horror of death from the skies rained down upon a helpless civilian population in the midst of their everyday lives, are brought to us via our TV and computer screens. On 16 November 1943, citizens of Rjukan became yet more victims to the bombers' scourge. They were not expecting this, nor were they prepared. How can civilians huddling in their homes ever truly be prepared? Their deaths were to be collateral damage, 'spillage' in the dreadful contemporary phrase. Just a minor calculation in the fearful arithmetic of war.

Even their home Government in Exile didn't know what was coming. I have not been able to trace any civilian record of this event, but during the Axis bombing spree of 1941, North Shields, a port on the Tyne, suffered one of the worst single losses when a random bomb from a lone aircraft struck a civilian shelter and killed 107 people. It serves to describe the horror. James Robert Ward witnessed the aftermath, searching for the bodies of his family:

> The last time I saw Mrs Gibson and Ethel was in the old Wash House on the corner of Church Way and Saville Street … I was looking at their faces, but they could not look at mine; they were dead, victims of the bombing of Wilkinson's Air Raid Shelter. The Wash House was utilised for a mortuary. I was in there with my brother Bill, looking for three of our family, they were there with many more we knew.
>
> My brother Bill and I went through the Wash House to identify our dead; we managed to get Bill's wife Lily out and their eldest daughter Lily. As we went into the mortuary, the attendant offered us a cigarette, Bill took one, I don't smoke. The attendant had a cigarette case with a lighter attached, the first that I had seen. He could not keep his hands still as he tried to light Bill's cigarette. This man had the worst job in Shields at that moment. Bill steadied his hand, and he took us around.
>
> The children and small adults were on the top bunks and the big people on the bottom bunks. It was like walking down Memory Lane, there was everybody I knew by name or sight, Mr Gibson and Ethel, the Broans from Tyne Street …
>
> I was going to go around the back of the bunks, when the attendant said, 'Don't go around there, there is nothing there for you.' He was wrong. I did go around and found my sister Lizzie. She had a black eye and I think that she had been pregnant at the time. The back of the shelter was like

> a butcher's shop, all of the corpses were wrapped in burlap, that kind of muslin that the butchers used. Bill had found Ann, Lizzie's daughter, but could not find Maureen, his own daughter. We climbed up the ladders to examine each child. One did look like her, but the attendant showed me three chalk crosses on the side if the bunk. 'Three people had identified this child as their own.' It was so hard to tell the difference.[1]

On that autumn morning in late 1943, a force of 300 US bombers lifted off from bases scattered across rural East Anglia, an area which hadn't witnessed this much military activity since Cromwell. The planes were Flying Fortresses[2] and Liberators[3] all belonging to US Eighth Air Force[4]. The Norwegian Government in Exile hadn't even been consulted. Some of the bombers peeled off to strike targets at Stavanger and Oslo, primarily as a diversion whilst the greater number – 162 aircaft – headed directly for Vemork. WWII bombing raids were almost always imprecise and the potential for civilian losses was always high. Men like Arthur 'Bomber'

B-17 Flying Fortress.

Harris were not likely to apologise – his avowed aim, when hitting German cities, was to cause maximum devastation and loss of life. This was total war.

But these weren't Germans, Norwegians were our allies, victims in every sense. The main strike was scheduled for lunchtime when it was hoped fewer workers were on the premises. Approaching in seven distinct formations, flying in a south westerly direction the giant planes, silver in the clear autumn air, hurtled up the narrow Rjukan Valley. This was a vision of *Ragnarök*. Noise blasted out by each plane was huge, drumming in a furious crescendo, real 'Twilight of the Gods' with the power to rain death and destruction on the monster's lair – if they could just hit the damn place.

They were early, 20 minutes before schedule. To his great credit the flotilla commander ordered the whole mission to turn around and take a second run at the correct time. If he hadn't, many more would undoubtedly have died. In so doing, he exposed his crews to the attentions of the Luftwaffe. Around 700 bombs (29.5 tons worth, a mix of 1,000lb and 500lb), were dropped from a height of 14,000 feet at precisely 12.12 hours.[5] Of those, 600 failed to hit their target. The raid lasted 45 minutes.

Add to the roar of the aero engines the scorching blast of explosions, ripping buildings apart, solidity and shelter turned into shrapnel and the tomb. Dust rolling in vast clouds, the stink of burst drains, the scream of pain and shock and disbelief. Exhaust from those planes in front obscured the view of those behind. The valley was a narrow shooting gallery and most who followed in behind the first wave had no idea of their actual target – thirteen planes turned back as they couldn't find a mark in the murk, but another dozen were drafted in from the subsidiary raids to splatter another 118 500lb bombs. Two of the bombers went down with a loss of 22 crew.[6]

Though the roof deck of the hydro plant and the penstocks to the rear had suffered hits, the heavy water high concentration cells, the key target, had escaped virtually undamaged, only 60 kilograms of deuterium oxide had been vapourised. In fact, only eighteen bombs hit the factory, while the power station took only four hits. For a while the flimsy link of the suspension bridge was down, half a dozen adjacent properties were flattened. Far worse, a crammed air raid shelter was obliterated. With a final flourish of awful irony, a lone skier, miles off, was blasted by a random detonation; the arithmetic of war can be fearfully capricious. In all, 22 ordinary people lost their lives. The Norwegian Government in Exile were both shocked and furious.

Why had all this become necessary? The *Gunnerside* team had seamlessly executed the sharpest of surgical strikes, they had excised the Nazis' capacity to manufacture heavy water and, in so doing, had surely stopped Hitler's attempt to build a nuclear bomb. The raid had taken no casualties, nor inflicted any, no civilians had been hurt. Von Falkenhorst, Terboven and the whole apparatus of oppression had been left bewildered and utterly humiliated. Their cack-handed efforts to catch the Commandos had foundered in a frozen ocean of blundering ineptitude.

Tronstad and SOE had expected this level of destruction would put the workings out of action for a good twelve months. But this was where the Germans, as ever, came into their own. Restoring capacity became their Holy Grail and organisation was what they did best. Terboven wanted the place back at full production within two months. Security was massively reinforced and loyal Norwegians who might be tempted to throw a spanner in the works would be aware of what fearful consequences would surely ensue. Einar Skinnarland, the Allies' local ace in the hole, was monitoring their moves and feared they might get the place started up by the end of May by diverting new gear intended for the next unit at Saheim back to Vemork. By July, Einar was dolefully convinced the enemy would be fully back in business by or before the end of August. If so, it was game on again in the nuclear arms race.

Having been once bitten, Von Falkenhorst wasn't about to offer the snake a second strike. Coil upon coil of new barbed wire entanglements sprang up, scores of mines were scattered like the hydra's teeth. All doors to the primary structure bar one were fully bricked up, windows were covered by steel mesh. Each garrison outpost was beefed up and defenders were now fully aware that any hint of a breach, or security failure, would merit a one-way bahnticket to the Eastern Front. By this time your average *landser* was definitely anxious to avoid such a posting, the opportunity to sacrifice yourself in the Russian mincer had lost any appeal it might have exerted two years previously.

SOE had scotched the snake, not killed it. Tronstad knew that a further precision raid couldn't now hope to succeed, and the alternatives all lacked appeal. Even a large-scale assault was highly doubtful, and the *Freshman* disaster was still in everybody's minds. Bomb the place? The location was difficult in the extreme, carpet bombing would have a limited chance of destroying the facility whilst certain to cause civilian casualties. Dams could be blown up[7] but this was another blunt instrument. Hitting the Mosvann dam would possibly do the job but would be bound to inflict significant loss of life, breaching that at Skarfoss could inundate the plant without killing civilians, but there were no guarantees.

A kamikaze was considered, a brave or very reckless loyal Norwegian to undertake a suicide mission from within the plant, accepting he/she had no chance of survival and that, inevitably, there would be reprisals against family and community. This was asking a great deal and currently, if unsurprisingly, there was no shortlist of volunteers. Seemingly, an impasse. Confusingly for Tronstad, the Germans, despite their fixation to get the works operational, didn't appear to have twigged that the conventional 'utilisation' method of production was being superseded by the more advanced 'combustion' method. He was wary of this, suspecting the Nazis were further ahead in the race than they really were. In this devilish quest for Armageddon, it was wisest to assume the worst.

Bombing became an increasingly attractive option, if for no better reason than there appeared to be no other. The USAAF was keen on the mailed fist. As the air war against Germany mounted in tempo, vastly exceeding anything the Luftwaffe had inflicted on Britain, the huge resources available to Allied bomber fleets meant that such a blow was entirely feasible. The term 'precision bombing' even with today's capacities can be very close to wishful thinking. Besides, Norsk Hydro, as Tronstad was keen to point out (quite rightly), was a very important Norwegian industrial asset. Even in wartime the plant was producing 85,000 tons of fertiliser annually.[8] Norwegians were suffering from the economic costs of the occupation as well, food imports had all but dried up and their fishing fleets had been forcibly corralled.

As early as 20 August, SOE were ruminating on the bombing option. Tronstad was dead against it, so he and his Government in Exile were simply marginalised, in fact excluded. Norwegian domestic interests were just not far enough up the pecking order of priorities. This seems wholly reprehensible, given the level of sacrifice Norwegians were already making and how brilliantly successful their Commandos had proved.[9] It's not easy at this remove of time and with the inestimable benefit of hindsight to see where the Allies were coming from. Nobody, certainly neither Gubbins nor Wilson wanted to inflict damage on a loyal, brave and oppressed ally, but the Domesday spectre of the bomb haunted them all. After four hard and bloody years of war, clawing towards a pathway to victory after so many defeats, fighting the most monstrous tyranny and then to risk losing all, was just too great a gamble. Norway must bleed.

For Leif Tronstad and his Government in Exile, this was both a catastrophe and a massive slap in the face. Their territory had been bombed, national infrastructure affected, with citizens killed. Their position was an invidious one. Here they were, guests of the Allies, sometimes useful but never holding a seat at top table, to be manipulated according to operational needs. They seethed with impotent rage yet ever mindful of the need not to overly antagonise otherwise obliging landlords. Their protest was a model of diplomatic restraint whilst not pulling any punches. The Norwegians stressed how fully they had cooperated until now and how completely they understood the needs of operational necessity. They hardly needed to add that the great success of *Gunnerside* was due to the work of *their* Norwegian Commandos.

They also went on to imply that such intense cooperation and willing sacrifice should not be taken for granted, they were allies in the fullest sense and more than keen to pull their weight. After all, it was *their* people beneath the jackboot. At the same time, they served notice, albeit in a reasoned manner, that as the tide of war was clearly turning, in fact had turned, in the Allies' favour, the Government in Exile had to look at new priorities – safeguarding the country's infrastructure and industrial base from deliberate Axis sabotage (or, by way of sub-text, anyone else's). Finally, and most persuasively, the Norwegians pointed out that besides being unsanctioned by them, the raid was grossly disproportionate in terms of the weight of the hammer being deployed, and equally ineffective in terms of its outcome. Fair points all.

To labour said points the Government in Exile went on to state they would have been happy to both sanction and coordinate a further Commando raid, and clearly February's success had demonstrated how effective this tactic could be. It was clear from all the evidence, and this was not inconsiderable, that the location, construction and very nature of the plant pretty much ensured a sledgehammer approach was doomed from the outset:

> The bombing of the Norsk Hydro works has made the deepest impression both on the great and loyal population in Norway and on the Norwegians fighting with their allies on the other battlefronts ... Messages received from Norway testify to the full to the bewilderment and dismay prevailing there. When one knows the enthusiasm with which other offensive actions against targets in Norway have been greeted, these reactions must be considered very seriously ... In making the above observations, the Norwegian Government must express their most profound regret at the events which took place.[10]

What was particularly galling was the fact that Leif Tronstad, of all people, had been deliberately kept in the dark. This was manifestly unjust as without his efforts, the threat would possibly never have been revealed in the first place, nor the Commando sabotage planned so effectively. It must have felt like a gesture of contempt. His own response was perhaps more cutting than that of his government.

Rather icily (and who can blame him), he confirmed that he'd already had discussions with USAAF officers and highlighted the virtual absurdity of trying to counter the threat with a deluge of 1,000-lb bombs. This was the moment when he had advanced his far more realistic proposal to bomb the Skarfoss dam. That went unheeded. He also reminded everyone that the Allies had previously agreed not to take further offensive action without bringing him onboard. He concluded his damning riposte by suggesting that the only outcome of the bombing would be that the Nazis would in fact take fright and ship their entire precious stash of heavy water back to the safety of the Fatherland.

He would have the grim satisfaction of being proved entirely correct.

Hitler might be losing the war, ranting and raving at his generals for all the mistakes he'd made, and he might, in more rational moments, have concluded the race for the bomb was likely over, but he wasn't about to give up. *Swallow* (Skinnarland) was on the case as ever and began to pick up murmurings before year's end, hints the Axis would indeed ship their entire supply out to Germany. Skinnarland was soon able to confirm that production at Vemork would cease and that Norwegian Quislings would take over responsibility for what would be mothballed.

By the latter part of January '44, this had changed from probable to a dead cert. We now know that Germany was dropping far too far behind in the race, but Allied intelligence at that juncture did not. It was still game on and the fact the Germans were going to go to such efforts to shift their stocks rather suggested they were still very much on track. This shipment had to be stopped and this time likely consequences in terms of reprisals simply had to be accepted. This wasn't cruelty or indifference; it was simply fact. Einar Skinnarland, quoting from his contacts at the plant was able on 7 February to confirm the move was likely to take place within a week.

There was therefore little or no scope for meticulous reconnaissance and planning, factors which had contributed so much to the successes of *Grouse/Swallow* and *Gunnerside*. This time however, the Norwegian government was consulted and, albeit with understandable reluctance, gave its consent in the sure and certain knowledge that any success would likely come at a price, possibly a very high one. Tronstad was fully in the loop this time, the lesson of the bombers' failure appears to have been grasped, but he was left in no doubt that the job was to be seen through regardless of civilian losses. He had no choice but to agree and, like all decent men, would do so with a heavy heart. Total war is just that, humanity is an early casualty.

Surely this was mission impossible: the route chosen and means the Germans would use to move their stocks of heavy water would be decided by them (even if options were limited), opportunities for effective and lasting sabotage were so much more limited. It was no use mounting an attack unless the product was wholly destroyed or irrecoverable. SOE would, at the very best, get one chance and one chance only. If that failed there'd not be another. Security would be present throughout at the highest levels possible.

Von Falkenhorst would be fully aware that the allies would discover his intent and that they'd try their utmost to frustrate him. Any assault began to look very much like a suicide mission. The odds appeared very heavily stacked. Most of *Grouse/Gunnerside* operatives were scattered or re-deployed, nigh-on impossible to re-unite within the tiny time window. Only two 'assets' remained available in the area, Einar Skinnarland and Knut Haukelid. 'Those were serious, burdened, rushing days, and the holes in the dyke called for all the fingers in our hands.'[11]

Even the intrepid Haukelid was daunted:

> A one man attack on Vemork was out of the question. All high concentrations and all concentrations down to

> 1% (which is a very high concentration), were drawn into different drums, 157 electrolysis tubes were tied together and filled 49 big drums. The large number of units made any sabotage extremely difficult. In a sabotage action, time is a limiting factor and one man, or even a small group, would have little chance to attach a charge to all these points and everything was very well guarded.[12]

Bonzo even had more personal worries at this stage. A letter had reached him from his wife in Sweden asking for a separation. He had begged three weeks leave to sort out his domestic matters and this had been granted. He'd not seen her for three years and was given the space he needed. Meeting up in Stockholm, the pair agreed on a divorce, and he returned to duty with a heavy heart. Happily, they were later reconciled.

On 9 February, Haukelid sent a message via Einar Skinnarland listing the difficulties and intimated they felt their best chance, indeed their only one, would be to sink the ferry transporting the drums. Failing that, they might try to derail the train into Svelgfoss Gorge.[13] They cautioned that such actions would almost inevitably spark reprisals.

Now they joined by a 26-year-old engineer from Vemork, Rolf Sorlie. Sorlie had form, fluent in German, having been trained at Leipzig. He'd previously joined Milorg despite a rare physical disability which resulted in constant spasming of the muscles in his hands and feet. The gang of three resolved that he and Haukelid should carry out a careful recce in Rjukan and then construct a detailed plan. Sorlie had no military training and received a very quick crash course – very different to the patient, gruelling months the Commandos had experienced in Scotland. He did, however, know the route from their Nilsbu hut to Rjukan. Helped by a couple of hefty farm hands to hump weapons, kit and explosives they set up temporary HQ.

Time might be of the essence, but Haukelid was too seasoned a campaigner to take silly risks. Sorlie might not be experienced but his intelligence gathering skills were clearly instinctive. The vital drums were labelled 'potash lye' by way of misinformation and a further two full companies of SS had been drafted in with a pair of spotter planes now based on the shores of Lake Tinnsjo. These were constantly flying over the surrounding terrain to pick up on any activity: 'The Germans knew from their D/F stations that we were somewhere in the mountains. Earlier they had said, "We know they are there. We can't find them, but we shall see they don't do any harm."'[14] Von Falkenhorst wasn't going to give away the keys to his pandora's box if he could avoid it.

Had he fully appreciated the deadly resolve of the handful of brave men planning to spoil his party, he'd have been even more worried. By careful marches, they stole ever closer to Rjukan, choosing a much closer hut above the town, well concealed in a wooded defile; so close they could hear the Germans practising their weapons drill 500 yards distant. By now, their lock-picking and discreet breaking and entering techniques were well perfected. They became nocturnal, sneaking down into the town for reconnaissance and contact, then climbing back up to their sequestered cabin to sleep during the day: 'I gradually became so accustomed to the ¼-mile journey up and down the hillside that I suggested to Rolf that we should get season tickets for the trip!'[15] The distance wasn't far, but the ascent was extremely steep, up 3,000 feet (914 metres), often over sheet ice, a gruelling three-hour trek.

Haukelid had both eyes and ears at the plant, including Gunnar Syverstad, Alf Larsen and Kjell Nielsen. The latter controlled transport in and out, Syverstad was involved in production and kept the Commandos well informed. Once these contacts became aware that Haukelid was planning to hit the ferry, they became understandably alarmed over the inevitable prospect of civilian losses. There was also of course fear that some ghastly nuclear detonation might be the broader consequence if the Germans retained their supply of heavy water. Though the Norwegian technicians doubted the Nazis were anywhere close to developing this capacity.

Mindful of these consequences, Haukelid and Skinnarland worked up a set of coded messages involving fish: a message referring to ten kilos meant the operation could be aborted but five meant it had to go ahead regardless. Einar sent a message to London expressing the Norwegian engineers' doubts as to real German progress. But the reply was unequivocal: 'It is thought very important that the heavy water shall be destroyed,' adding somewhat complacently, 'Hope it can be done without too disastrous results.'[16] Five kilos.

There was nothing to do but get on with it. The team, Haukelid, Syverstad, Sorlie, Neilsen and Larsen, conferred on their available options. Choices were indeed very limited. A direct attack on the plant itself was immediately ruled out, at least a full platoon would be needed, and they could only muster a half-section. Next up would be to have a go at rail transport between Vemork and Rjukan or between the town itself and the ferry landing at Mel. Handily, Norsk Hydro's explosives store – a whopping 4,500 lbs (2,040 kilos) of dynamite – was housed adjacent to the line.

Not necessarily as tempting as it seemed. Any blast would have to occur at that precise moment the train was passing, a detonator on the rails could sort this but both the store and the line would naturally be guarded. So, any guards on the dump would have to be eliminated, the break-in effected, and detonator placed. Timing would be everything, with the added spanner in the works that the Germans might well sent a single loco on ahead of the train as a sacrificial goat. Altogether too many unknowns.

The third alternative was to simply blow up the entire train at Svelgfoss, which lay between Tinnoset at Notodden. Civilian casualties with the added risk of contamination from ammonia drums, also on the train, would inevitably be high. The last and only truly viable option was to sink the ferry and ensure she went down at the deepest point on Lake Tinnsjo. Once the drums plummeted down 1,300 feet (400 metres), they'd be irrecoverable. All were uneasy about the inevitable toll of innocent lives that would result, but their orders were crystal clear.

As Haukelid points out in his testimony, it is one thing, as an officer, to risk the lives of your men – they are soldiers, it's in the job description – but to plot an action which will kill non-combatants, those who are also your countrymen and women you're fighting to liberate, is a wholly different matter. In this case, Haukelid was the only officer in-theatre. His instructions from London, approved by his own Government in Exile, were exact. The show must go on. We rightly laud the physical courage of these men, which is without equal and yet this decision, this very lonely and terrible decision, exhibits another kind of courage and one which certainly requires the same level of fortitude.

Circumspect as ever, Haukelid did a further recce, crossing on the ferry he intended to destroy armed with a concealed Sten and brace of grenades. He found it took 20 minutes to arrive at the spot known as Hasleviken, where the cold waters were at their blackest and deepest, the abyss where, if he got it right, he would very literally sink Adolf's chances of ever winning this vital race. Next, the device. A clock and detonators were required. Rolf Sorlie, who never lacked resourcefulness, managed to acquire some in Rjukan.

The Germans seem to have picked up on the vibe, getting steadily twitchier as the big day drew near. Their garrison now outnumbered the indigenous inhabitants. They were stopping and searching with fanatical thoroughness. Battered, weather-beaten and in distinctly shop-worn gear, Haukelid would be unable to escape their curiosity and watchfulness. Happily, there was also a concert, one of those 'aren't we wonderful and cultural' affairs that occupiers like to put on to show how reasonable they are. Ironically, this gave Haukelid his cue. Smartly attired in a borrowed blue lounge suit and kitted out with a violin case (shades of Al Capone), he strolled undetected through the streets and yes, he had a Sten in the case.

Next job was to test the efficacy of their newly acquired detonators. Outside their hut some of these were laid on some spare baulks of timber, connected to an alarm clock. The timer was set to go off when they were due to rise from their daily slumbers that evening. They certainly didn't oversleep, the bang, correctly timed, sent them reeling from their bunks to face a non-existent attack. Game on.

What made matters more complex was the fact Milorg had mobilised following a wrong intimation that an Allied invasion was imminent so they should disperse into the hills and make ready. With the organisation in Rjukan thus disrupted, contact was fractured. Whilst the enemy failed to apprehend anyone involved, they had netted a coach-full of youngsters making their way out of town. The lads fessed up, either innocently or disingenuously, that they'd seen this run for the hills and were fleeing themselves, as they believed the Allies must be invading. Pretty thin perhaps but the Germans swallowed their tale and let them go.

What Haukelid needed was transport, ideally a private car. One of the group's contacts was a resistance man who'd earlier fled to Sweden but returned to his home town as he found this exile disagreeable. Haukelid intended to remove him thence again when he himself got out – his presence was a clear risk to Milorg. This individual did, however, produce a suitable vehicle complete with driver. The clock was, in every sense ticking Neilsen had managed to delay movement aboard the ferry until next Sunday, when hopefully there'd be fewer passengers. To bolster his alibi, Neilsen was to undergo surgery for appendicitis and had booked his operation for the Monday morning so he'd be under anaesthetic when the Gestapo might be interested in talking to him!

The next and most important trick was to get the device onboard unnoticed.

On Saturday, the ball or rather the train began to roll, when the barrels filled with heavy water were loaded onto rolling stock and taken down to Rjukan Station. Nielsen's intelligence confirmed they'd be shipped across the lake on Sunday. Sorlie had a pal who intended to board the ferry that fateful Sabbath, but it was felt too dangerous to warn him; the traveller had just bought a ticket in the lottery of war.

Syverstad's mother was also due to travel on SF *Hydro* and Haukelid advised simply locking the old lady in a closet! Nerves were taut, not just the saboteurs, Alf Larsen was aware of increasing Axis telephone traffic, nobody wanted to be responsible for another disaster. As an added precaution the Germans had chosen to divide their precious cargo after Notodden when half would continue by rail whilst the remainder was conveyed on the roads.

For a brief spell the team pondered the possibility of an attack on the train while it stood at Rjukan station. Fleetingly tempting but this implied that each of the 50 drums would have to be wired up individually, it was highly unlikely the massive German security presence would be that negligent. In the hour before midnight on Saturday 19 February, Haukelid and Sorlie crossed the bridge over the River Maan. The railway trucks were brightly lit and crawling with sentries. At 08.00 the train would roll down to the ferry terminal, ready for departure two hours later.

Their own acquired transport proved somewhat less reliable. Haukelid had sourced some fuel, but the engine refused to play, it took a nerve-wracking extra hour to get the damn thing started. It wasn't till nearly 01.00 that they stuttered into Mel, pulling up around ¾ of a mile from the quay and warned the driver to be ready and to ensure the motor was running. If they didn't make it back inside two hours, he was to drive off regardless. Alf Larsen was in the car, with instructions to make a run for Sweden. With him gone, the Germans would have lost the last senior engineer at Vemork with working knowledge of the processes.

Dead of stark, cold winter's night, 'armed with Sten guns, pistols and hand grenades, we crept past Mel station and down towards the ferry. The bitterly cold night set everything creaking and crackling; the ice on the road snapped sharply as we went over it. When we came out on the bridge by the ferry station, there was as much noise as if a whole company was on the march.'[17] Yet all was quiet in that crystal air, no crunch of hob-nailed boots, in fact no Huns at all: 'Was it possible that the Germans had omitted to place a guard at the weakest point in the whole route of transport?'[18]

Bonzo went forward alone to recce, guns of the rest covering: 'Hearing voices in the crew's quarters, forward I stole to the companion and listened. There must be a party going on down there and a game of poker.'[19] The other two boarded and the trio of saboteurs crept stealthily across the deck, descended to the third class lounge, and discovered the hatch which opened up to the bilges. This was where the device was to be planted. Then, suddenly, they had to dive for cover as the figure of the boat's watchman was framed in the door. This was tricky, but the man proved to be loyal and was only too willing to believe their story of being on the run. Obligingly, he lifted the hatch for them. The local Milorg resister now animatedly engaged the pliant watchman while Bonzo and Rolf lowered their backpacks into the space and went to work: 'It was an anxious job, and it took time. The charge and the wire had to be connected; then the detonators had to be connected to the wire and the ignition mechanism. Everything had to be put together and properly laid. It was cramped and uncomfortable down there under the deck and about a foot of water was standing in the bilge.'[20]

Then it was done: '19 pounds [9 kilograms] of high explosive laid in the form of a sausage'. It was placed well forward so that when it detonated the stern of the craft would tilt upwards, rudder useless, propeller spinning in the air. Ideally, the angle would cause those laden railway trucks (the deck was fitted with tracks for easy roll on, roll off), to plunge into the icy depths. The explosion would blast a hole of, say, one square metre in the bow. The lake was very narrow so SF *Hydro* had to go down fast before she could limp to the shore and be beached. Haukelid had thought about this long and hard, the effect of the sabotage had to be instant and fatal. There was still hope most of the passengers could be saved.

To be doubly sure the team used two alarm clocks, the hammer would short-circuit the current as it commenced striking, and these were fixed up on one of the ship's structural ribs with a wire feeding down to the explosive. Bonzo had relied on the train being ten minutes late, so he set the alarm for 10.45 when he calculated *Hydro* would be crossing the very deepest part of Lake Tinnsjo. Making the final, very fine adjustments was as nerve-wracking than the rest of the operation, or worse. The clock wasn't really intended for such work and there was never more than a gap of a third of an inch between hammer and alarm – one third of an inch from a very loud and final exit, mission unaccomplished.

Perhaps the watchman, still busily engaged in chat with the local resister, had shared in the onboard conviviality but he never became suspicious, even when the trio announced they needed to get back to Rjukan to pick up more gear, but they'd be back in time for the crossing next morning. Haukelid and Sorlie were exercised by the risk they were exposing this crewman to. Should they take him with them on some pretext? They didn't dare, even knowing he'd be a prime suspect in the eyes of the Gestapo, assuming he survived the sinking. Instead, they shook hands and left, thanking him for his contribution. Perhaps he was surprised, but he didn't know just how much he had in fact contributed.

Back to the car – and just in time, their two hours were nearly up. Rolf Sorlie was to join Einar Skinnarland in the hills while Haukelid headed for Sweden once again. Near Kongsberg the driver dropped them off, he had to get back to Rjukan before dawn to avoid attracting suspicion. Bonzo and Larsen kept moving on skis until they reached the town's station. This was tricky as well, no Norwegian was permitted to travel more than 15 miles (24 kilometres) from home and, sheer bad luck, a German troop train was just pulling in. Feldgrau covered the platform and Larsen had to lock himself in the gents – he was known to the Gestapo in Rjukan and given the sensitive nature of his job he wasn't allowed to travel out of the town at all. The moment passed and the Gestapo agent onboard went on with his journey. Finally, the coast was clear, and the pair were able to board a train for Oslo: 'Halfway, I looked at my watch; it was quarter to 11.'[21] Zero hour. He and Larsen didn't talk much after that.

Rolf Sorlie had given Bonzo a contact on Heroya[22] from where the heavy water would be shipped on to the Fatherland if the sabotage should fail. Unbeknown to them, SOE had already sent in a team to mount a second attempt with Royal navy submarines skulking in the Skagerrak if nothing else worked. Haukelid knew nothing for sure until he picked up a newspaper from a stand in Oslo next day: 'Railway Ferry *Hydro* sunk in the Tinnsjo'. Mission accomplished.[23] 'Larsen and I continued our escape. We left the train outside Oslo. The Oslo Central Station was a bad place where the Germans used to search all passengers, the city was warier than it had been when Poulsson and Helberg met there in the aftermath of the earlier raid. At the same time all German higher officials again concentrated at Rjukan.'[24] As well they might, this was going to take some explaining. Meanwhile, as the Nazis fumed and sorted their alibis, the pair of saboteurs got clear away.

Another who was thirsting for news was, of course, Einar Skinnarland, who darted forth from his mountain lair to pick up his copy of *Fritt Folk*, the Quisling broadsheet. Their banner headline also proclaimed the loss of SF *Hydro*, confirming that the precious cargo of heavy water locked in heavy railway trucks had slid forward as the bows tilted and plummeted into the very deepest part of the lake, exactly as Bonzo had planned it. Einar's transmission to London gave terse news of success but did also mention civilian losses. SOE's operatives in Sweden were able to elaborate, reporting the deaths of eighteen people, fourteen locals and four occupiers. A young couple from Notodden and their three-year-old daughter had been lost. It was the less affluent passengers in the cheap seats below who suffered most.[25]

An 80-year-old farmer, Gudleik Persgaard, was one of those who ran to their boats and rowed furiously to the rescue. He made two trips out and saved several lives. One of these survivors was the boat's captain, Erling Sorensen, who reported that he heard an explosion while on his bridge at around 10.30. *Hydro* heeled over to port and was soon lying on her side. Sorensen walked along the now horizontal flank of his stricken vessel, stripped off his coat (while forgetting to remove his heavy sea boots) and swam for it. Once clear, he turned to see her slide under prow first, exposed propeller still thrashing full throttle.

As a postscript, even though the bulk of her precious cargo crashed down into the depths, four barrels later bobbed to the surface. None of the Allied observers could quite work this one out, until further intelligence from within the plant explained that these containers were at best only half full and the heavy water in them was of a very weak concentration, virtually useless to the Germans. Their precious and much cherished cargo was a total write-off.

Bonzo had just beaten Hitler and there'd be no German bomb. Now, for the Nazis as they gazed into the black depths of lake Tinnsjo, it was their own abyss they were staring into.

10

NORWEGIAN SPRING SUNSHINE

'So, get up into those mountains and get back every one of those men. Dead or alive. And if it turns out they're Norwegians, I want 100 hostages rounded up and shot. I don't want even a rat left alive up there.'

Heroes of Telemark

Remember before God those men and woman of the European Resistance Movement who were secretly trained in Beaulieu to fight their lonely battle against Hitler's Germany and who, before entering Nazi-occupied territory here found some measure of the peace for which they fought.

Plaque in Beaulieu Abbey Cloisters unveiled by General Gubbins in 1969

Claus Helberg's war was very far from over:

Another Mission In the night of 4 October 1944, nine soldiers parachuted at Ugleflott. One of them was Major Leif Tronstad. My original team leader, another team leader, another team member, and myself were also among these men. Einar Skinnarland stood on the ground with the landing lights. Our assignment at this time was to assist the resistance

Nuclear weapon of the 'Fat Man' type, the kind detonated over Nagasaki. The bomb was 60 inches in diameter and 128 inches long. It weighed about 10,000 pounds and had a yield equivalent to about 20,000 tons of high explosive. (Naval History and Heritage Command)

force in Upper Telemark with weapons and training. It was of decisive importance for the economy of Norway after the war that the valuable industry in Telemark should remain intact. The Germans could be desperate enough to destroy the factories when they withdrew. The resistance would help to prevent this. Our headquarters was established in Skinn Valley by Mos Lake, with Tronstad as the chief.

During the winter, we requested and received more than 100 'drops'. Almost every moonlit night, large Stirling planes came with guns, ammunition, food, uniforms and radio equipment. We had radio stations in every district, and we communicated among ourselves and with England. We established so-called cells, and the men in them were fully trained militarily. They were the elite troops. Training with weapons was given at night in the farm areas and in the towns. We had field manoeuvres in the mountains on Sundays. The Germans were more alert than ever, and their raids were part of our daily routine. We regarded the Hardanger Plateau almost as

> a recovered area. The blockade was still on, but the Germans had no chance of maintaining effective control there.
>
> Then came 8 May. All the groups mobilized. On the mountain farms at the Mos Beach and Rauland, in every town and farm area, men and young boys dressed in uniforms. They had been prepared to defend their country. Fortunately, actions were not necessary. A long, liberating breath was let out by all Norwegian lungs. Professor Tronstad did not live to see that day. On 11 March, he and another team member were shot by a traitor close to our headquarters. It was a great loss for Norway. Tronstad was an important scientist, and in no way could we afford to lose him. The people in the mountain farms around the Mos Beach mourned Tronstad deeply. Never had one person been so loved by the people on those isolated farms. All through the winter, he had lived with them. They appreciated his inexhaustible spirit and optimism, and they especially were fond of him for his considerate and friendly nature and for the warmth of his heart. For years to come, the memory of Professor Leif Tronstad will live on among the Mos Beach people.[1]

Norwegian high command had kept Leif Tronstad out of the direct fight. He was desperate to go but remained simply too valuable. His contacts within the UK from his pre-war days in research and academia were many and varied. At the same time, he was able to stay in touch with leading scientists still stuck in the Norwegian Institute of Technology, Harald Wergeland, a leading physicist and Njal Hole, chemical engineer and nuclear physicist (also active in the resistance). Tronstad had a genius for organisation, topping off his formidable skills set. Having begun in England as part of the MoD, he moved to the Norwegian High Command where from 1943 he headed up his own section in department 4 (FO IV). Originally, this sub-unit had been set up by a Norwegian Air Force Colonel, Bjarne Oen, and Tronstad focused on operations aimed at industry and shipping as well as taking responsibility for *Kompani Linge*. As the war neared its end, his role shifted more towards the protection of Norwegian trade and business from last ditch Axis sabotage attempts.

This would be complex. That critical relationship between SOE and the Norwegian Resistance, Milorg, was never going to be an easy one. Churchill had ordained the creation of the former to foment discord and carry out sabotage throughout the whole of Hitler's wide European satrapy, whilst the latter was inevitably a home-grown response to the invasion of their native heath. As we saw earlier, Norway was effectively taken by surprise and occupied before the populace could even think of fighting back.

Whilst the country offers much mountainous terrain ideal for *maquis*-type insurgency, it is sparsely populated and with relatively few large urban centres. The Nazi repression was so sudden, so overwhelming and so utterly demoralising that people were essentially paralysed. The Germans knew a thing or two about repression, they were practised, thorough and – as Norwegians very quickly came to appreciate – not bound by any rule of law or even basic humanity. Such ruthlessness can produce results and they had, at least at the outset, a willing 5th column in Quisling's faction.

Yet time was not, as it seemed in the beginning, on their side. In 1940, the sheer dazzling mass and it has to be admitted, brilliance of Blitzkrieg swept all before it. The Allies, powerless to defend Poland, chased firstly from Norway then hustled humiliatingly out of France and Belgium, seemed inept and impotent. France, always billed as the premier military power, was flattened in just over a month. Nazis in gleaming boots stomped along the Champs-Elysées and wandered at will through Frogner Park in Oslo. If the Allies' combined might couldn't stop Hitler, what chance did Norway have with a population of only three million, barely equal to London or Paris?

Resistance would come, it began haltingly and in a small way but soon began to emerge as something more focused and better organised. Plenty were still prepared or even willing to collaborate, and the Government in Exile was fearful that communist elements might influence or hijack the movement. Milorg was wary if not fearful both of SIS and SOE, large-scale sabotage operations might be box office for Winston but would inevitably lead to wholesale and savage reprisals. Terror was a standard tactic in the nazi toolkit and the greater the affront, the more vicious the riposte.

Yet it soon became clear the Nazi war machine wasn't invincible, Britain remained defiant, the pendulum in North Africa continued to swing and when on 22 June 1941 Hitler unleashed *Barbarossa*, initially sweeping all before him, mud, frost and ill-preparedness stopped him just before the gates of Moscow. This was now truly a world war; Britain was no longer isolated once the US joined in. Russia, even if only ever an ally of convenience, could not be easily subdued. Even those Norwegians who had accepted the imposition of the jackboot began, in many cases, to resent its unbending weight.

Resistance throughout occupied Europe was never constant. Most countries were riven by political differences which affected their objectives, Gaullists and communists squabbled in France, communists and monarchists feuded in Greece (indeed this would lead to a vicious civil war). Milorg, too, had its factions, inevitably so, even if these were never as deep or harmful. Resistance in Norway did fall under the control of the High Command of the exiled government, which proved a broad church, permitting local organisations a largely free hand.

In September 1941, the first incident of mass unrest occurred when workers in the capital went on strike. This was far from armed resistance, but Terboven reacted as though it was, mass arrests, torture and killings. He tried to confiscate all wireless sets, but many were hidden and the BBC remained a lifeline to the free world, and the underground press began to flower.

Hitler's increasing fear that the Allies might, as indeed Churchill favoured, decide to bust open Fortress Europe by landing first in Norway prompted the dispatch of a further 150,000 *Wehrmacht* personnel there during the first half of 1942 – the realization of *Festung Norwegen*. In so doing he weakened the Atlantic coast of France and effectively imprisoned his own forces in a part of the world from which they would be totally unable to influence the Allies main strategy.

We've already looked at the stellar achievement of the Shetland Bus and the hundreds of small craft which escaped from Norway providing a readymade fleet for SIS/SOE. By the autumn of 1942 as *Grouse/Swallow*, *Freshman* and *Gunnerside* were gearing up, SIS already had sixteen listening posts in place. SOE would undertake twenty-two sabotage attempts and infiltrate more than a dozen clandestine radio operators. As this increasing hum of activity swelled, so did the will to resist. Between 1940 and 1941, some 3,300 loyal Norwegians escaped to carry on the fight; roughly 10 per cent were killed or captured. Freedom is never cheaply bought.

Though relations between SOE and the RAF had initially been difficult, increasing Allied air superiority and more planes being constructed meant that air supply could not only function but would become crucial to the survival of mushrooming resistance groups. During November 1944 for example, SOE and Milorg were supplied with a hefty inventory of kit arriving by both sea and air; 53 tons of stores: 1,628 rifles, 322 carbines, 120 Brens, 860 Stens, 1,014 Mills bombs and eight bazookas, a formidable arsenal.

From 1943, as the sands of fate began to run through Hitler's increasingly trembling fingers, plans were underway for an invasion of Norway. The Germans achieved a relatively easy victory against home forces in 1940 and they were now entrenched and in large numbers. Against such likely opposition, Norway began to look like a very tough nut to crack. Milorg were part of the plan but potentially part of the problem. Resisters after these years of brutish thuggery were raring to rise up and attack the occupiers. Several instances in France, however, showed how dangerous this could be. However enthusiastic the fighters might be, they'd never be any kind of match for the *Wehrmacht* in a conventional fight.

Happily, Milorg got the message and made it clear to its members that patient preparation must be order of the day and that precipitate action could only bring disaster and the unspeakable horrors of mass and indiscriminate reprisal. As the Germans saw victory slipping so quickly away and aware the Allies would soon be dismembering their hoped for 1,000-year Reich, the tempo of their cruelty only quickened.

Arrests, deportations, killings increased. A sense of irony wasn't a noticeable feature of the Nazi occupiers, but they wondered why their efforts to recruit workers to serve the Fatherland and soldiers to fight on the Eastern Front met with such little success. Some did go into that great mincer, few came back. Ideology can be a great and potent weapon, but its very introversion and inflexibility can be its downfall. By 1944, as the Allies came ashore in Normandy, Terboven, Rediess and the whole of the occupation apparatus were, like Macbeth 'in blood / Stepped in so far, that, should I wade no more, / Returning were as tedious as go o'er.' (Act III.4. 136–8.)

As voluntary recruitment had failed Terboven, in 1944 he ratcheted up the pressure by issuing an edict that all males born between 1921 and 1923 should muster at Oslo's labour exchange for compulsory service. This, known in France as *Service de Travail Obligatoire* ('STO'), had proved a huge boost to recruiting for the maquis as thousands of young men fled into the hills. In Norway, SOE agents in the capital destroyed the machines processing the mobilization cards and, as in France, many likely conscripts simply headed for the nearest high ground. Next up, the *Reichkommissar* attempted to exert control through the issue of new ration cards, those who didn't play ball would go hungry, but Milorg neatly intercepted the train ferrying the new documents.

> Soldiers, Sailors and Airmen of the Allied Expeditionary Force! You are about to embark upon the Great Crusade, toward which we have striven these many months. The eyes of the world are upon you. The hopes and prayers of liberty-loving people everywhere march with you. In company with our brave Allies and brothers-in-arms on other Fronts, you will bring about the destruction of the German war machine, the elimination of Nazi tyranny over the oppressed peoples of Europe, and security for ourselves in a free world.'
>
> Eisenhower's Order of the Day.

D-Day was a natural inspiration for Norwegians, yearning to be free of the Nazi yoke and while Eisenhower pushed for the raising of an increased army of 30,000 resisters, he tempered this with caution, warning that while deliverance was creeping closer, the Battle for France had to be decided first. After the Normandy campaign was finally won in August and a vast caravanserai of fleeing Germans annihilated in the Falaise Pocket, the Axis grip seemed to unravel. Allied tanks roamed at will through Northern France and Belgium. In the east Operation *Bagration* gathered huge momentum. But then the pace slowed, Operation *Market Garden*, the attempt to end the war with a single dynamic thrust, failed. Against all odds the *Wehrmacht* rallied and attacked in the Ardennes as the Americans bled in the hell of the Hürtgen Forest. Deliverance postponed.

Meanwhile, Leif Tronstad would, with tragic irony, get his way and take his crusade directly to the reviled enemy. Tronstad's mission dovetailed a quartet of linked operations under the banner of *Sunshine*: *Moonlight, Starlight* and *Lamplight*. Given the track record of those who took part, including Poulsson, Helberg and Kjelstrup, this sounds like a recipe for mayhem on a grand scale. (The others were Norman Lind, Herluf Nygaard, L. Bronn and Gunnar Syverstad.) But that wasn't the case. As mentioned, this was more about safeguarding installations than blowing them up. The plan was to carefully monitor what might happen to the power stations in Telemark with Buskerud, encompassing Norsk Hydro plants in Rjukan and Notodden. *Moonlight*, centred on Rjukan, was led by Poulsson, *Starlight* under Kjelstrup focused on the Numedal, northeast of Rjukan. Herluf Nygaard (an experienced resister and post-war professional soldier), headed up *Lamplight*, concentrating around Notodden.

A Norwegian spring was coming, the most magical time of year there when winter's long icy grip, cloaked in never ending darkness, retreats. Days at last get longer, new light appears like a beacon, melting snow creates glittering, impromptu waterfalls. Spring brings hope and a sudden profusion of arctic wildflowers: willow, moss campion, sedum, draba, svalbard, poppy, Arctic chickweed, snow buttercup, all spreading like a carpet over ground so long hard as iron. On the evening of 11 March 1945, Tronstad was, true to his dearest wish, active in the field, back in his beloved Norway. This was where he wanted to be and by now with war's end imminent, the vast crushing tentacles of the Nazi War machine were beginning to lose their grip and recede, the seemingly impregnable fortress collapsing, falling in on itself. That wasn't necessarily all good news. The Nazis had a habit of destroying what they couldn't pilfer.

Of course, Tronstad didn't need to be at the sharp end, he'd more than done his bit. His mission that night was relatively mundane, or it seemed to be. With two members of his team, Gunnar Syverstad and Jon Landsverk, he was interrogating a Quisling sheriff, Torgeir Longnvik. Tronstad had two objectives: firstly, to secure information on other fellow travellers in the region and secondly, to ensure that no information on resistance activities reached the ears of the Gestapo. The enemy's power might be on the ebb, but his teeth were still sharp. The nervous official's life was potentially on the line, but Tronstad was not a cruel or vindictive man and decided simply to corral his captive for the remainder of hostilities – he'd be under Haukelid's watchful care at Bamesbu.

Their present location was quite remote, a two-room shieling close to Lake Mos, and the indefatigable Einar Skinnarland had been instrumental in netting this influential Quisling. Questioning over, the resisters were making ready to leave the hut when suddenly, the door burst open and Johans Longnvik, Torgeir's brother, burst through firing wildly. Gunnar Syverstad was struck in the head and as he spun back, pushed Landsverk clear of the fusillade. Tronstad leapt for Johans but stopped his next two bullets. Crumpling up, he was probably finished off by blows from a rifle butt wielded by Torgeir, who joined in his brother's frantic assault. They fled leaving the two dazed survivors locked inside. Leif Tronstad was dead, Syverstad was clearly on the way out and it took a while for the dazed Landsverk to break out of the cabin and dash off to find Skinnarland.

Einar Skinnarland might have been terribly shaken at this sudden and malicious twist of fate, but his professionalism kicked in. Both brothers would soon be back with their German chums. All papers and other evidence were gathered up and burnt. There was little that could be done for the desperately wounded Syverstad, who soon succumbed. The two survivors dragged their dead comrades down to the frozen lake, hacked holes in the ice and sent both fatalities to their watery graves. There was no time for ceremony. Needless to say, the Germans soon uncovered the bodies which, post inspection and photographing, were doused in gasoline and incinerated. The war might have been nearly over, but Nazi mores didn't change.

Poulsson assumed overall command of the *Sunshine* operations. Leif Tronstad never did see his wife and children again.

'What Rewards for These Heroic Men?' Winston asked, though he, better than most, understood they had never sought glory for its own sake.[2] Very few of them were, in any real sense – despite their national service in 1940 – professional soldiers, though several did continue with military careers after the war. Like so many hundreds of thousands of others they'd say they just 'did their bit'. It was just that their bit was rather more than most. Very much more than most. Both Ronneberg and Stromsheim confirmed in post-war interviews that the full import of their task wasn't immediately clear to them. 'We weren't certain what the heavy water in Vemork was intended used for,' said Joachim Rønneberg and Berger Stromsheim.

> We were not informed that the Germans wanted to use heavy water to develop an atom bomb. But before we left England, we were told that our action had the highest priority. But we were in no way aware of the large bearing it could have on the outcome of the war. Joachim Rønneberg (70) and Birger Stromsheim (78) from Ålesund, on this weekend, were back at Vemork together with those other saboteurs from Company

> Linge. This group carried out the blasting of the heavy water plant at Vemork during the night on 28 February 1943. It is the first time since the war that the group of men is gathered at that place.[3]

Group leader Joachim Rønneberg:

> For the Allies, it was important that the sabotage in 1943 was successful. We had an assignment (a mission) to accomplish. When at war one does not ask its purpose. Our task was to carry out the blasting. That was for us to accomplish. We knew that 40 British Commandos had lost their lives in an attempt a couple of months earlier, and that Vemork stood in danger of being bombed by Allied planes if our attempt failed. With that, many civilian lives could be lost. But it was after the war, we first got to know what heavy water was to be used for.

Together with Birger Stromsheim he carried the responsibility for placing the explosive charges which destroyed the heavy water plant at Vemork.[4]

Joachim Ronneberg was born into Norwegian aristocracy, a distinguished line. *Gunnerside* wasn't the end of his active service. He went on to lead other teams of Commandos in the field including Operation *Fieldfare* in Sunnmore, laying the ground for a series of blows against German logistics throughout the Romsdal Valley. At the start of 1945 when the great Axis war machine was staring its own *Gotterdammerung* in the face, he commanded a three-man group which blitzed a significant transport link, the Stuguflat Railway Bridge, deploying a huge 290-lb (130-kilogram) blast of plastic explosive that put the span firmly out of action for a vital three weeks – time the Germans could not afford to lose. In addition to his DSO, he received a clutch of decorations from his grateful country and others: St. Olav's Medal with Oak Branch, Defence Medal 1940–1945, the Haakon VIII 70th Anniversary Medal, the US Medal of Freedom with Silver Palm, *Légion d'honneur* and *Croix de Guerre*. In 2015 a statue of him, crafted by Hakon Anton Fageras, was unveiled in his home town of Alesund by Princess Astrid.

Post-war, he married a crafts teacher Liv Foldal and embarked on his new career as a broadcaster, beginning in 1948 with NRK based in Alesund, remaining in the media until his retirement in 1988. As ever, he remained modest about his pivotal time with SOE. It took a long time for Ronneberg to talk openly of his wartime experiences but 50 years later he did for a local Norwegian TV station under the banner, 'War-saboteur reveals':

> For 50 years the war-saboteur Joachim Rønneberg (82) has avoided film projects about his war-efforts. He now makes an exception, hoping the rest of us will learn.
>
> 'I agreed to take part in this documentary project, because I want to tell the younger generations about the knowledge which was forced upon us (by the war). Joachim Rønneberg is the man who directed one of the most successful and famous Second World War raids. Destroying the heavy water factory at Vemork in Telemark, set production back by a half a year and it contributed to the delay of the German effort to produce an atom bomb ...
>
> This is far from the first time the Vemork-raid has been the subject of a movie. The story about the sabotage action has been told several times as a documentary and feature film. But the man who has had detailed knowledge about it, has so far always refused to contribute. 'I've been afraid of not being able to approve of the final result,' he explained. That's why he did not portray himself when the movie *The Fight Over Heavy Water* was made in 1948, as many of his friends and colleagues in the resistance group did. Later he experienced a Hollywood portrayal of himself in *Heroes of Telemark* with Kirk Douglas in the role of Rønneberg ... Joachim Rønneberg's scepticism about taking part in the film projects has created a myth that he does not want to talk about his war experiences. This is not so. The last years he has been giving lectures to thousands of students.

> 'The first couple of years it was perhaps a little difficult to talk about the things I had experienced. As time has passed, I've come to realize that people have a right to ask, and they are entitled to be answered.' For Joachim Rønneberg the fight began in earnest as he was leaving for Scotland in 1941. He wanted to join the Navy, but after a couple of days met Martin Linge of the close to legendary Linge Company.
>
> There he became a fast-rising star – among the British too. 'Rønneberg was one of the most outstanding men we had. He was well-balanced, intelligent and awfully hardy,' narrator Colonel Charles Hampton said in the film programme. Just before Christmas 1942, Second Lieutenant Rønneberg, designated as the explosive material expert, was put in charge of the action against the heavy water plant at Vemork. At the end of February the following year, the sabotage action was accomplished-without loss of life. Joachim Rønneberg speaks matter-of-factly about his contributions.
>
> He prefers to place the emphasis on his fellow saboteurs: Birger Stromsheim, Knut Haukelid, Fredrik Kayser and Kasper Idland to mention a few. Or telegraph operator Olav Aasæther. He also becomes eager and enthusiastic when the subject changes to present day defence-politics. "'It is frightening that we no longer have a proper military defence in today's Norway,' Rønneberg reflects ... 'We must not forget history. It can happen again.'[5]

Post retirement, Ronneberg was involved in a conservation project which restored his own HQ – the *Fieldfare* cabin – re-imagined as a museum and heritage centre. He lectured regularly in schools and was a commentator on defence affairs, highly critical of the perceived failures of the Norwegian Government to have heeded the all-too-obvious lesson of 1940: Always be ready! He was lauded for cementing strong relations with the USA and a stalwart member of the *Linge* Club, a veterans' association which survived until autumn 2007. Ronneberg was handed a presentation Union Flag at an SOE commemoration in London in April 2013. Joachim Holmboe Ronneberg died on 21 October 2018, aged 99.

Jens-Anton Poulsson returned to Norway with Tronstad in 1944 as part of the *Sunshine* team. As noted, there were nine SOE agents dropped by parachute over Ugleflott in Ovre that October. The irrepressible Helberg was their radio operator. The overall mission, as we saw, was to establish fighting resistance units in three regions, Rjukan, Notodden and Nore (Milorg District 16 or 'D-16'). Poulsson was responsible for the first of these and built up a strong company of 300 local volunteers, supplied by air. After Tronstad's death, he assumed overall command of D-16. Like his *Gunnerside* mates, he not only gained his DSO but the Norwegian Defence medal with Rosette, Haakon VI's 70th Anniversary Medal, the St. Olav's Medal with Oak Branch, and the War Cross.

Post-war he remained in the Norwegian Armed Forces, holding a succession of high-profile postings, rising to a full colonelcy. He commanded the King's Guard, a very prestigious appointment, 1961–1965. When he died in February 2010, aged 91, he was treated to what was almost a state funeral, with King Harald V in attendance, a fitting tribute to a very remarkable and courageous Norwegian.

Knut Magne Haughland remained, together with Skinnarland, marooned on the Hardanger for two months after the successful sabotage at Vemork, training Einar in the varied arts of a clandestine radio operator. Then he left to go back to Oslo so he might bring on a fresh crop of resisters, subsequently returning to the UK but only for supplies. By November 1943, he was back in Norway, parachuting in with another celebrated patriot Gunnar Sonsteby.[6] Haugland, like Helberg, possessed the devil's own luck; just as well. The Gestapo caught up with him in Kongsberg, but he managed to give them the slip for a second time. On 1 April 1944 he experienced another very close shave when a German radio direction finding unit pinned his location down to Oslo's maternity hospital. He just got clear in time, made it back to England and remained there for the rest of the war.

After hostilities ended and Norway was freed, he, like Poulsson, stayed on in the Norwegian military. Two years after the war, he volunteered as a member of the epic Kon-Tiki expedition. His amazing skills as a radio operator paid enormous dividends, invaluable to the safety of the explorers. In 1951, he married Ingeborg Prestholdt and put on uniform again, serving in Germany 1948–1949, three years later transferring to the Norwegian Air Force. As an airman, he headed up electronic surveillance in North Norway, a key job in the most intense years of the Cold War, then at its height.

Leaving the air force as a lieutenant-colonel, he moved to curatorial posts as director of Norway's Resistance Museum and then curator of Kon-Tiki from 1991. Always, like Wilson in England, a keen scout (six of the seven man-crew of Thor Heyerdahl's great experimental craft were scouts and Haughland was the last survivor). As a tribute to his seriously varied CV in both peace and war, he garnered a significant collection of decorations and medals besides his DSO & MM: The Royal Norwegian Order of St. Olav, the War Cross with Sword (and bar), the Defence Medal (with three stars), US Medal of Freedom with Bronze Palm, *Legion d'honneur*, *Croix de Guerre avec Palme*, Knight 1st class of the Order of Vasa, Knight 1st class of the Order of Dannebrog, Knight of the Icelandic Order of the Falcon, Cross 1st class of the Federal Republic of Germany's Order of Merit, Commander of the Belgian Order of Leopold, Companion of the Order of the Crown of Thailand, Officer of the Iranian Order of the Lion and the Sun, Grand Decoration of Honour in Gold for Services to the Republic of Austria, together with the Peruvian Order of Merit for Distinguished Service.

While he was no fan of the movie *Heroes of Telemark*, he did play himself in the 1950 *Kon-Tiki* documentary. Knut Haughland received yet more awards during his later years. In 1988, as recognition for his services to history and heritage, he was confirmed as a Knight 1st class of the Order of St. Olav. He died aged 92 on Christmas Day 2009.

Kasper Idland returned to Norway in 1944 where, fighting with group Vestige 4, he remained active in the Egersund area.

The team's primary objectives were disruption of shipping and commerce, though, as the war progressed, preservation of assets began to eclipse sabotage. The resisters sank no ships, but during January 1945 they did blitz an AT camp at Bjerskreim. As the hated oppressor's defeat finally loomed closer, Idland shifted his efforts into the Setesdal area, to set up a haven or *Varg*, controlled from a mountain *saeter* at Langelona where recruits could be re-supplied from the air. By 8 May, VE Day, the group was still only a nucleus.

With peace restored, Idland moved back to Stavanger until 1955, when he emigrated to Farmingdale in New York State. He died there in May 1968, aged only 49. As well as his MC, Kasper Idland received the St. Olav's Medal with Oak Branch, War Medal, Defence Medal 1940–1945 with Rosette, Haakon VII's 70th Anniversary Medal, *Legion d'honneur*, *Croix de Guerre*, US Medal of Freedom with Bronze Palm. Years after his death, in 1995, he was honoured with a memorial set up in his home community of Figgio.

Einar Skinnarland had a very active war. An engineer by training and profession, he was never a professional soldier, in a sense his job at Norks Hydro was just happenstance. But his buccaneering escape aboard the hijacked steamer *Galtesund* was merely a beginning. After being dropped back onto the Hardanger Plateau followed by the high adventure of the raid itself, he remained very much committed. By the time peace was declared, he had been promoted to captain and was in charge of resistance forces in Rjukan and Northern Telemark. Britain awarded him a DCM and he received a Norwegian War Cross with Sword, US Medal of Freedom, plus *Croix de Guerre avec Palme*. His 2003 obituary (he was 84) read:

> Mr. Skinnarland was an engineer by training and a patriot by inclination. He joined the Norwegian resistance and became a British agent, operating in his occupied homeland under threat of certain torture and death if captured. His early messages were decoded by the famous cryptographer Leo Marks, who passed them on to Winston Churchill's war

> cabinet. At first, Mr Marks was frustrated by what appeared to be coding errors but then he stumbled upon the importance of Mr. Skinnarland's work.
>
> 'It was only when I spotted two words tucked away in the last line that I realized which of us was the chronic invalid,' Mr. Marks wrote in his bestselling 1998 memoir, *Between Silk and Cyanide*. 'The words were "heavy water".' They had been distilled by Morse mutilation into *heaxy woter*. I wondered what heavy water was.' The Norsk Hydro facility at Vemork produced hydrogen for use as ammonia in fertilizer. One of the by-products was deuterium oxide, known as heavy water, which was believed vital to the preparation of uranium-235 for use in an atomic bomb. The Special Operations Executive (SOE) was desperate to know about Germany's interest in atomic research and summoned their agent to Britain. Of course, they were unable to provide transport. Mr Skinnarland was introduced to Odd Starheim, a like-minded resistance fighter with a flair for daring. The two got along famously. 'It was sabotage at first sight,' Mr. Marks wrote.[7]

Three years after Germany's defeat, he actually played himself on screen in a Norwegian movie, *Kampen om Tungvannet* ('The Fight over Heavy Water'). It was US actor Brook Williams who played Skinnarland in *Heroes of Telemark*. In 1965 he emigrated to Canada and for the rest of his distinguished career worked as a civil engineer on large-scale works:

> Among his many projects were the Detroit Dam in Oregon; the massive Bhakra Dam in Punjab, India, which on its completion in 1963 was the fifth-highest dam in the world at 226 metres; and the Robert Moses Power Dam on the St. Lawrence Seaway. Other projects included consulting work on the Trans-Alaska Pipeline and the offshore oil platforms of Hibernia. As well, Mr. Skinnarland found himself on the other side of the dam builders when hired as the principal technical and business consultant of the Grand Council of

> the Crees of Quebec during negotiations over the James Bay Project in the 1970s.
>
> As project manager of the relocation of Fort George, Mr. Skinnarland was responsible for moving a village of 2,373 at the mouth of La Grande Rivière and relocating it to a new site upstream. The village is now known as Chisasibi, which is Cree for 'great river'. Mr. Skinnarland suffered a stroke in March 2001. He died of a brain haemorrhage in Toronto on Dec. 5 after a series of accidental falls. He leaves his wife Trudi (née Gans), who was born in Fairview, Alta.; five adult children; six grandchildren; and his sister Solveig. Mr. Skinnarland was uncomfortable in accepting tributes for his wartime actions. Instead, he always noted the heroism shown by those unknown Norwegian civilians who lost their lives while aiding the resistance.[8]

Knut Haukelid wrote up his adventures and his pivotal role in sinking SF *Hydro* in his memoir *Skis Against the Atom,* on which I have drawn heavily. Whilst the necessity of this act was generally acknowledged, lingering resentment over the civilian death toll still simmered and Haukelid received death threats until the very end of his life. Aside from his British awards, he was honoured with the Norwegian War Cross, firstly with sword (1944) and then, three years after, with Two Swords. He remained serving with Norway's Armed Forces, graduating from the Military Academy in 1948. He was promoted to major in the Telemark Infantry Regiment, made up to Lieutenant-Colonel in 1959 and spent the final years of his service commanding the Oslo Home Guard, retiring in 1974.

A decade later on the 40th anniversary of the raid, those ageing survivors of *Kompani Linge* were feted by the US Ambassador at his official residence in Oslo. The following autumn on 18 October 1985 Haukelid was feted again at the 2nd Annual Scandinavian-American Hall of Fame Dinner in Minot, North Dakota. He died, aged 83, at home in Oslo on 08 March 1994.

Klaus Helberg had undoubtedly expended most of his nine lives (probably more). He, perhaps more than any, symbolises the kind of Buchan-esque adventurer so beloved of spy fiction, a man whose astonishing courage, fortitude, resilience and sheer bloody mindedness exemplifies the unquenchable spirit of resistance. His contribution was very widely recognised: Knight 1st Class of the Order of St. Olav (1986), St. Olav's Medal With Oak Branch, King's Medal of Merit in Gold, Defence Medal 1940–1945 (with rosette), King Haakon VII's 70th Commemorative, Knight of the Order of the Dannebrog (Denmark), *Legion d'honneur*, *Croix de Guerre*, US Medal of Freedom with Bronze Palm, October Revolution 80th Anniversary (Soviet Union, 2000).

Having – surely against the odds – survived the war, he went back to guiding, employed by the Norwegian Mountain Touring Association until he finally retired in 1982. Even then, he stayed on as a volunteer. His clients included numerous members of the Norwegian Royal family, including Queen Sonja and Margarethe II of Denmark. Claus Helberg died in March 2003 in his 84th year.

Fredrik Kayser was another who had a seriously busy war. *Gunnerside* wasn't his first mission. He'd taken part in Operation *Anklet* (the raid on the Lofoten Islands) late in 1941. Together with Kasper Idland and another *Kompani Linge* graduate, Kjell Endressen, he had planned another operation, codenamed *Woodcock,* a daring plan to parachute onto the Nevlandsheia Plateau in Gjesdal (part of Rogaland county) and from there assault the Sola Air Station (still a Norwegian Air Force base today). Bad weather frustrated their first attempt and subsequent related operational failures finally scuppered the plan.

Following the Telemark Raid, Kayser got back to the UK and, if *Gunnerside* hadn't been exciting enough, re-trained on one-man midget subs, generally viewed as shortcuts to Valhalla. In 1944, he with three others were deployed to attack Axis vessels in Måløy Harbour on the island of Vågsøy. It didn't work, largely because the locals became aware, and the submariners had to make a run for it overland. As resilient as ever, he found he had to spend a whole day immersed in a reeking bog to throw off German hunters. Next year, as the noose began to tighten around the neck of Hitler's Thousand Year Reich, he was back for more.

This time the plan was to set up a series of secure havens, or *Vargs*, five in all, as refuges for those resisters forced to run but also as in-country training centres to harden the spear of armed insurrection. Kayser with fellow *Linge* Commando Lieutenant Severin Synnes (wonderfully piratical) were put ashore by sub (HNoMS *Vigra*) at Masfjorden in October '44. Their mission was to set up safe area *Bjorn West*, Matre in Vestland County. Things at the outset went smoothly enough, but German patrols found them and battle raged from 28 April to 3 May. The resisters not only survived but gave a very good account of themselves.

Kayser. Too. survived but left the army after his country was free and became a successful entrepreneur, a devout Christian who also dabbled in local politics, broadly conservative though without direct affiliation. He was, amongst numerous decorations, awarded the St. Olav's Medal (and Bar) and the *Legion d'honneur*. Though reluctant to talk about his wartime exploits he did, nonetheless, appear as himself in the movie *Kampen om Tungvannet* (1948). Thereafter, he pretty much clammed up and it wasn't till the 1990s that he finally collaborated with author Kjell Harald Lunde, whose account *The Saboteur* was published in 1997. Three years previously he had agreed to appear in a Norwegian Broadcasting Corporation documentary on the fight for *Bjorn West*. He had retired to Agotnes where he died aged 90 in February 2009. His son Johan played Fredrik in another TV version.

Hans Storhaug got clear back to England after the success of *Gunnerside* but returned to Norway that October when he participated in Operation *Grebe*. This went tragically wrong from the start. Half the team of six died on landing as they were dropped into a frozen lake, leaving Storhaug and the remaining two to tab (again, in his case), across to Sweden and safety. Undeterred, he and the other survivors tried again next March – Operation *Grebe Red* – and successfully operated in the Osterdalen region till the war ended. *Grebe* and *Grebe Red* were SOE operations, established independently from Milorg. The agents were to build up a force of saboteurs and keep close watch on the Roros and Dovre Railways, part of that vital thread of rail communications that facilitated German troop and logistical movements with a view to demolitions should the need arise. Storhaug also played himself in *Kampen om Tungvannet*. His decorations included

St. Olav's Medal with Oak Branch, Haakon VII's Anniversary Medal, *Legion d'honneur* and the US Medal of Freedom. He died in June 1995 at the age of 80.

Arne Kjelstrup stayed behind after the raid, working alongside Haukelid. Both were extremely proactive in recruiting throughout Western Telemark for Milorg's appropriately named, *Bonzo* Group. Finally, he made it out through Sweden and returned safely to his adopted home where he undertook yet further training in demolitions. He was with Leif Tronstad for Operation *Sunshine* and commanded the Nore Section – Operation *Starlight*. By VE Day he'd mobilised a battalion of fighters 900-strong. Kjelstrup, who lived post-war in Baerum, (a municipality which forms an extended suburb of Oslo) died, aged 80, on 5 May 1995, missing the 50th Anniversary of the liberation he'd fought so bravely for by only a few days. Aside from his British medals he was decorated with the St. Olav's Medal with Oak Branch, Norwegian Defence Medal 1940–1945, *Legion d'honneur* and *Croix de Guerre*.

Ronneberg described Birger Stromsheim as 'beyond doubt the best member of the party'. Given the men who he was serving with, that's very high praise indeed. On his return from the *Gunnerside* mission, he received his St. Olav's medal with Oak Branch from the hand of King Haakon VII himself, with the award ceremony fittingly held up at STS 26, where so much training and preparation had taken place. He had also earn the Defence Medal with Rosette and King Haakin VI's 70th Anniversary Medal. Like the rest of the team, his war hadn't finished in 1943, and he returned to Norway to establish a guerrilla base in Romsdal Valley. Also like so many of the team, he was none too impressed with *Heroes of Telemark* – 'too glamorous'! Stromsheim died in November 2012 in Oslo, aged 101.

It was the Russians who first appeared as 'liberators' when they chased the Germans through Finland, their erstwhile ally. As ever, the Nazis left a trail of terror and waste in their wake, and still Milorg waited for Norway's western allies. It wasn't until the Fuhrer's suicide and the remnants of his regime was handed to Admiral Donitz that Germany finally surrendered unconditionally. A senior British officer accepted the formal submission of the Axis forces in Norway. Right on cue, an army of fighters emerged from the shadows to disarm the occupiers, patriotic flags like spring flowers blossomed through cities, towns and villages. The long shadow was lifted.

Meanwhile, Knut Lier-Hansen who had escaped from the area following the sinking of SF *Hydro*, on 8 May, together with a couple of others, was handed the job of tracking down and arresting *Reichkommissar fur die Besetzen Norwegischen Gebiete* Josef Terboven. The wheel was coming around full circle. The posse caught up with their quarry at Skaugum, official residence of the Crown Prince, 12 miles (19 kilometres) southwest of Oslo, from where the *gauleiter* had lorded over his temporary subjects. His grand fiefdom was reduced to a few fortified acres and though his remaining sentries resisted, Terboven knew the game was up and the noose being made ready. He opted for an explosive finale, deploying 50 kilograms of dynamite. General Wilhelm Rediess died the same day in the same redoubt, but he shot himself prior to detonation. Lier-Hansen lived to be 91 but garnered few plaudits for his resistance work. Like Haukelid, he continued to receive death threats over the sinking of SF *Hydro*.

What these men, who were in every true sense heroic, achieved, what they endured, would hardly be credible in any work of fiction. Their mission and its extraordinary success would help to define our modern understanding of Special Forces operations. Any who considered Commando style ops as boy's own swashbuckling and little more, were confounded. A team of ten men in-country had not only landed and survived under the noses of a generally vigilant enemy, but they had also penetrated their great redoubt, wreaked circumscribed havoc and got clear away. Then one of them returned and sank the Nazis' precious supply of heavy water, in such a way it could never be salvaged. Mass bomber assault failed to do this and resulted in a higher civilian death toll.

Grouse/Swallow and *Gunnerside* must rank as the most successful Commando operations of the Second World War.

We've looked at *Musketoon* and others and there were the big raids like St. Nazaire (Operation *Chariot*), the ill-fated big raids in North Africa (Operation *Agreement*) prior to El Alamein and the doomed epic of the Cockleshell Heroes (Operation *Frankton*), which, whilst brilliant, daring and certainly heroic, was carried out at heavy cost and to little overall effect. Operation *Anthropoid*, the assassination of Reinhardt Heydrich in Prague, resulted in the deaths of the whole team and ghastly mass reprisal against the civilian population.

Modest as ever, the *Gunnerside* team felt their successes could be attributed to 'careful planning, good luck, and co-operation'. These are the reasons they give for the success of the action, notably accomplished without loss of life.

> The sabotage squad, consisting of nine people, made it to the heavy water plant without being detected by German guards. No crisis occurred until the blasting party were to enter the room containing the heavy water tanks, while the others kept guard outside. By mistake, members of the blasting party lost each other. Two of them, among them Rønneberg, got into the plant through a cable intake and began to place the explosives.
>
> Suddenly the window in the room, was broken in. We thought it was the Germans. Fredrik Kayser – who was our lookout – almost started to fire. But the controlled fear during such situations has the effect of a safety-valve. It makes a man think twice before the action is carried out. This saved Birger Stromsheim' s life, as he was the one who broke the window. Fortunately, it took the German guards some time to show a sign of reaction after the explosion inside the heavy water plant. This gave the Linge guys an unexpected advantage during the pull out which kept them alive.
>
> Birger Stromsheim and Joachim Rønneberg express that the sabotage action was in no way an exceptional war operation. 'Our action was not more special than others during the war. Besides, we had additional detailed information about the

target before we went in. That was rarely the case with similar operations.'[9]

These young Norwegians had demonstrated what highly trained and motivated raiders with the right tools and equipment could achieve. Post-1949, as the North Atlantic Treaty Organisation (NATO) settled into the largely passive routine of the Cold War, with the prospect of a major interstate Armageddon looming daily, there appeared to be little requirement for Special Forces. The Cold War was a kind of dry repeat of the Western Front, both sides dug in and facing each other with little or no movement.

Warfare post-1945, tended to be a very different business. For half a century, industrial age armies faced each other across the North German Plain, each driven by fear of the other. Both possessed vast arsenals of nuclear weapons which assured that neither side could possibly win, 'mutually assured destruction' 'MAD' made it so. At the same time, Britain experienced the retreat from Empire, sparking asymmetric conflicts across the globe, Malaya, Kenya, Cyprus, Aden and Borneo. Simultaneously, there was scope for expeditionary warfare in Korea, Suez, the Falklands, two Gulf Wars and Afghanistan with thirty years of domestic counterinsurgency ('COIN') in Northern Ireland.

It is remarkable to think senior commanders after 1945 envisaged a continuation of Industrial warfare without the need for Special Forces. Today, we tend to view our Special Forces as the front line in any asymmetric conflict: 'The period 1946–1991 could well be defined as one of an overarching confrontation ... maintained by industrial structures, containing non-industrial conflicts, the parallel wars. It is within these conflicts that we see the first signs of the new paradigm, especially in the nature and objectives of the opponents, and in the constant adaptation of the existing means – the industrial military machines – to non-industrial conflicts ... It was the beginning of a trend we live with to this day.'[10]

Whilst the nature of the threat and the need for Special Forces has changed, in part at least, many of the core concepts underlying the creation of SOE still hold good. Gubbins raised his force to perform specific tasks in very particular environments. Success in one sector, as SOE several times discovered, did not guarantee a repeat elsewhere. Nowhere was success more evident than at Telemark. No other raid achieved so much at so little cost (not forgetting the *Freshman* disaster). Combined Ops and SOE learned how not to organise an attack. Success in war tends not to come cheaply. The Germans, bloody, ruthless and savage as they were, remained right until the end, formidable adversaries.[11]

It is axiomatic therefore that the formation and role of SOE shows that defining the job of Special Forces in the specific theatre from the outset is crucial. The 'KISS' principle ('keep it simple, stupid') is not one to be lightly overlooked. Expansive combined Commando-type operations such as Operation *Agreement* ended in disaster. This is a lesson which subsequent planners, of grandiose Operation *Eagle Claw*, for example – the 1980 attempt by the US to rescue hostages in Iran – failed to heed. This one also foundered, creating both a military and PR catastrophe. The lessons of both *Freshman* and *Gunnerside* weren't appreciated.

As General Kristoffersen comments on current theories of special operations:

> Looking at Operations *GUNNERSIDE* and *GROUSE* through the special operations lens, two different views may explain the successful outcome of the operations. Admiral William H. McRaven's SPEC OPS, assesses principles that lead to success for a small force in an offensive operation, whereas Dr. Robert G. Spulak's essay 'A Theory of Special Operations' focuses on the qualities of the personnel that are needed in order to succeed. These two positions complement each other and explain how Operations *GUNNERSIDE* and *GROUSE* achieved success.
>
> McRaven's six principles for special operations are: surprise, speed, purpose, security, repetition, and simplicity. From eight special operations case studies, McRaven derived that the key to success is 'a simple plan, carefully concealed, repeatedly and realistically rehearsed, and executed with surprise, speed, and purpose'. By following these principles,

SOF personnel are able to overcome the fictions of war if they also hold the 'moral factors of courage, intellect, boldness, and perseverance'. Spulak's theory is focused, as well, on overcoming friction in special operations, but he contends that the attributes of SOF personnel are the key to success rather than principles of execution ...

In Spulak's view, the three vital attributes, or qualities, of SOF personnel are: 'elite warriors, creativity, and flexibility'. Through extensive selection and training, along with prioritizing quality over quantity, SOF units may create teams with personnel who possess these attributes to a larger extent than personnel in regular units.

As a result, SOF teams may overcome friction to a larger degree than a conventional unit, which gives the SOF team a broader span of operational capabilities and fewer limitations. Together, McRaven's and Spulak's theories demonstrate how a small SOF team of specially selected and trained personnel may achieve success. Where a conventional unit may be limited by the specific challenges of an operation, SOF units with special skill sets may overcome these challenges and use them to their advantage (e.g., covert parachute insertion of a small team where the terrain limits movement of larger conventional forces).

Moreover, the limited size of a SOF unit may also benefit the operation, since smaller teams have a better chance to avoid enemy surfaces, operate covertly, and achieve the element of surprise ... At Vemork, Operation *FRESHMAN* experienced severe challenges due to the terrain and weather in Norway. Conversely, the SOF operations depended on the terrain and weather to get to the objective without being compromised by the numerically superior Gelman guard force. It was especially the SOF characteristic of certain access that proved to be vital for the success of SOE's operations at Vemork.[12]

It was these experiences and successes of SOE that signposted some important features for successor Special Forces. Suitably built or adapted kit, designed for heavy load carrying, rugged, reliable and not overly sophisticated, is essential. The failure of Axis forces to grasp this partly explains their own failures. Automatic weapons delivering a hail of fire way beyond the normal firepower of so small a group is frequently decisive in the vital opening seconds of a firefight. Early fire supremacy not only kills the enemy, but it also hastens their urge to disengage. It can also persuade them that they're facing a far superior force. Workable communications are a must. The brilliance of these Norwegian radio operators like Haughland and Skinnarland proved to be a crucial asset.

Any military unit depends upon the calibre of men it attracts. It has always been clear that you can't conscript Special Forces. They must, by definition, be volunteers. From those who come forward, only a small percentage will finally qualify. Current SAS selection is notoriously and rightly rigorous, owing a significant debt to SOE and such hardy perennials as Fairbairn and Sykes. None but the best will do. Gubbins and Wilson established this principle from the outset, using the trusted cadre of those he'd explored alongside before the war. They were all ordinary men, yet they were also magnificent men; resolute, resilient, courageous and bold.

Just to survive in the conditions they found themselves, expected to endure on the harsh tundra of the Hardanger Plateau, required each man have a steel core, officers doubly so. Their mission way, way behind enemy lines in the harshest environment on Earth imposed huge demands. Physical fitness and stamina had to be matched by emotional strength and motivation. There was no chain of command, no relief ever in sight, no backup, no re-supply. The squad were their own army. They carried whatever they needed. They had to tend their own and be prepared to see to any wounded, no chance of a scapegoat when things went wrong: 'Corporals' wars' in army jargon.

None of this has changed. For Special Forces to be able to operate successfully, only the bravest and the best can pass the test. Each man knows he can rely totally on his comrade. Officers have to muck in, to listen to the specialists whose combined skills determine unit effectiveness. There is no room for passengers and service in the field will quickly and cruelly expose those not up to the mark. From the Maccabees to Desert Storm, this hasn't changed, nor ever will. As much if not more than any other group it was these Norwegians who set the benchmark. They still do, probably always will.

We are the Pilgrims, Master,
We shall go always a
Little further:

It may be beyond the last blue
Mountain barred with snow,
Across that angry,
Or that glimmering sea.

LEST WE FORGET

GLOSSARY

AA	Anti-aircraft
Bivvy	Bivouac
GIGS	Chief of the Imperial General Staff
C-in-C	Commander in Chief
CO	Commanding officer
Combined Ops	Combined Operations Headquarters. Admiral Keyes was director 1940–1941 when he was replaced by Commodore Lord Louis Mountbatten, whose title was changed to Chief of Combined Operations – a post he held until September 1943 when he was replaced by major-General Robert ('Bob') Laycock.
CP	Command post
CWGC	Commonwealth War Graves Commission
DCM	Distinguished Conduct Medal
DSO	Distinguished Service Order
DUKW	amphibious vehicle
Flak	anti-aircraft fire
Gestapo	*Geheime Staatspolizei* (1933–1945); official secret police both inside the Reich and throughout all occupied territories
GOC	General Officer Commanding
HE	High explosive
Heer	The German Army 1935–1945

HQ	Headquarters
IO	Intelligence Officer
MAUD	Military Application of Uranium Detonation
MC	Military Cross
MG	Machine-gun
MiD	Mentioned in Dispatches
Milorg	'Miitaer Organisasjon' – the Norwegian Resistance
MM	Military Medal
MO	Medical Officer
MTB	Motor Torpedo Boat
NCO	Non-commissioned officer
O Group	Orders Group
OB West	*Oberkommando* West – Supreme Command West
OH	Official History
OKH	*Oberkommand des Heeres*
OKW	*Oberkommand der Wehrmacht*
OP	Observation Post
RA	Royal Artillery
RE	Royal Engineers
RMC	Royal Marine Commando
RSHA	*Reichssicherheitshauptamt;* Reich Security Main Office, run by Himmler as his dual fiefdom as both chief of German policing and head of the SS
RSM	Regimental Sergeant-Major
SA	*Sturmabteilung*; the 'Brownshirts' headed by Ernst Rohm until the infamous purge, 'The Night of the Long Knives', June/July 1934
SD	*Sicherheitsdienst*; from 1931 the intelligence agency for the 3rd Reich, the SS and Nazi Party
SIS	Secret Intelligence Service (MI6)
SIPO	*Sicherheitspolizei* – state political and criminal investigation agency, combining Gestapo and *Kriminalpolitzei* ('KRIPO'), 1936–1939, until combined with the RSHA

SOE	Special Operations Executive
SS *Schutzstaffel*;	Originally Hiler's personal bodyguard but expanded to become the official attack dogs of the Reich
STS	Special Training School
Tommy-Gun	Thompson sub-machine gun
USAAF	United States Army Air Force
W/T	Wireless telegraphy

APPENDICES

1. Proclamation to the People of Norway by the King of Norway and H.R. Nygaardsvold, Prime Minister, April 10th, 1940

The German Government has demanded that the king of Norway shall appoint a Government which enjoys the confidence of Germany, and which has been designated by the German Fuhrer. The king has been unable to submit to the demand, which would make Norway a dependency. No other government can have control here but that which has the confidence of the Norwegian people.

The Government, which has now stood for five years in control of the country in collaboration with the Storting [Norwegian Parliament], is still the only legal Norwegian Government.

This Government offered to resign when the German attack upon Norway started. But the Storting unanimously requested it to remain in office. And the Government remains.

It now turns to the whole Norwegian people and asks their assistance in upholding a rule of the country according to law – Norwegian Constitution, Norwegian liberty, Norwegian independence.

Germany has perpetrated against Norway one of those hideous acts of violence of which history has known too many instances. It has forced its way into the country with bombs and every other kind of destructive weapon. It has flagrantly and dishonourably violated all the rights of a small people, which merely wishes to live in peace.

The Norwegian government is convinced that this crime is condemned by the whole civilised world. But above all it is confident that the Norwegian nation will use all its energies to restore the liberty and independence which a foreign power has wished to destroy by force.

The outlook for Norway may seem dark at this moment, and the masters of force may succeed in destroying much. But the Government has the confident hope that a new and brighter future for the country will one day appear.

It therefore exhorts the whole Norwegian nation to cling fast to the Norwegian heritage of liberty in loyalty to the great ideas which have carried our country forward for centuries.

Long live our fatherland! Long live a free Norway!

JOHAN NYGAARDSVOLD

I associate myself completely with this appeal which the Government has addressed to the Norwegian people. I am convinced that I have the whole nation with me in the decision which has been taken.

HAAKON R.

2. *The* Freshman *Contingent*

'A' Apple

Pilot: Squadron leader A.B. Wilkinson (with Group Captain T.B. Cooper)

Glider

Pilot: Staff-Sergeant M.F. Strathdee with co-pilot Sergeant P. Doig (both from 1st Glider Pilot Regiment, Army Air Corps, Officer Commanding: Lieutenant D.A. Methven with Lance-Sergeant F. Healey, Corporal J.D. Cairncross, Lance Corporals W.M. Jackson & T.L Masters, Sappers J.F. Blackburn, F. Bonner, J.W. Walsh, T.W. White, J.G.V Hunter, E. Jacques, R. Norman, E.J. Smith plus drivers G. Simkins & P.P). Farrell.

'B' Baker
Pilot: Flight Lieutenant A. R. Parkinson (of the Royal Canadian Air Force); co-pilot Pilot Officer G.W. de Geney (RAF(VR)), crew; Flying Officer A.T.H. Haward (RAF(VR)), Flight Lieutenant A.E. Thomas (RAF(VR)), Sergeant J. Falconer, Flight Sergeant A. Buckton (RAF(VR)), Flight Sergeant G.M. Edwards (RAF(VR)).

Glider
Pilot: Pilot officer N.A. Davies (Royal Australian Air Force); co-pilot Pilot Officer H.J. Fraser (Royal Australian Air Force), Officer Commanding Lieutenant A.C. Allen with Lance Sergeant G. Knowles, Corporal J.G. Thomas, Lance Corporals F.W. Bray & A. Campbell, Sappers G.S. Williams, E.W Bailey, C.H. Grunday, H.J. Legate, T.W. Faulkner, H. Bevan, L. Smallman, J.M. Stephen plus drivers J.T.V. Belfield & E. Pendlebury.

On 19 July and 22 August 1945, all the flags in the Norwegian town of Stavanger flew at half-mast. In quiet and dignified ceremonies and with full military honours, men of 1 Airborne Divisional Engineers re-buried their killed and murdered comrades. Each coffin was draped with the Union Jack and covered with red roses from the people of Stavanger.

Source: Royal Engineers, Chatham Archive
Part 1. The Mission
In the summer of 1941, a scientific committee had reported to the Prime Minister, Winston Churchill, that there was a reasonable chance that the atomic bomb could be produced before the end of the war. After the project had received the blessing of the Chief-Of-Staff Committee the prime minister discussed the development of such a weapon with the President of the United States at the time of his visit to Washington in June 1942. The development of that joint Anglo-American weapon and the path to the dropping of an atomic bomb on Hiroshima is a separate story. The immediate problem was a likely German development of a similar weapon. Winston Churchill wrote:

> We knew what efforts the Germans were making to procure supplies of 'heavy water'- a sinister term, eerie,

> unnatural, which began to creep into our secret papers. What if the enemy should get an atomic bomb before we did! However sceptical one might feel about the assertion of scientists, much disputed among themselves and expressed in jargon incomprehensible to laymen, we could not run the mortal risk of being outstripped in this awful sphere.

From this strategic appraisal a number of possible tactics were considered. The plant had been built on a broad rock shelf above a steep ravine at the foot of Gausta Fjell. Most proposals, including bombing, were rejected as impractical. Ultimately, at the request of the Special Operations Executive (SOE), Operation *Freshman* was conceived. In simple terms, aided by pathfinder SOE Norwegians who were to be parachuted-in a few days earlier, two glider-loads of sappers were to land on a frozen lake close to the plant, attack the plant and disperse the stocks. At the same time they would damage the plant sufficiently to prevent further heavy water production, but not destroy the whole of the generating capacity on which much of that part of Norway relied for its electricity and fertiliser production. The troops were then to escape as best they could across snow-covered mountains to Sweden where, at the worst they would be interned, or more likely, repatriated. CRE 1 Air Div, Lieut-Colonel Mark Henniker, who was called upon to implement the plan was very unhappy at the possible outcome.

He raised four objections:

> Firstly, that it was illegal: he was persuaded that it was a legal operation and that it did not violate the Geneva Convention.
>
> Secondly, he offered the opinion that the Royal Air Force was not capable of getting the troops there. The RAF said that it could.
>
> Thirdly, he said that he did not believe that the results justified the high risk: he was told that he did not know enough about physics to be a judge of that.
>
> Fourthly, Henniker said that even if the Royal Air Force managed to land the troops in the right place, the troops

> were being asked to undertake an impossible task. He sought permission to lead it personally. The Divisional Commander, Major-General F.A.M. Browning, refused to allow him.

In the end, all of his objections were overruled and, in later years with all of the facts available to him, he himself believed that the government was right to insist on the attempt being made.

It was an operation that would haunt him for the rest of his life: indeed, it scarred everyone who was in any way connected with it.

With such a high-risk operation, planning took place in great secrecy. A cover story was devised around a fictional contest in Scotland with American Airborne Engineers for the prize of 'The Washington Cup'.

Elements of the Norwegian plant had to be destroyed, whatever the cost. The actual target was unknown to any of us other than the CRE and even he did not know the ultimate purpose of the raid.

Thirty high-calibre soldiers were selected from 9 Field Company RE (Airborne) and 261 Field Park Company RE (Airborne). All were familiar with gliders and were also qualified parachutists, as a decision was not taken until a late stage in planning on which method of transportation might prevail. After vigorous training in mountains and familiarisation with specialised power plants they left their units for Scotland. They never returned.

On 19 November 1942 the two gliders, tugged by Halifax aircraft, set off from Skitten airfield near Wick in Scotland. One tug and both gliders were reported as having crashed somewhere on the Western coast of Norway. The Field Security officer who covered the training and departure was full of praise for the sappers. All were aware that an abnormally high risk was involved yet went off with a cheerful courage.

On the morning of 22 November, newspapers carried a German report:

> On the night of November 19-20th two British bombers each towing one glider flew into Southern Norway. One bomber and both gliders were forced to land. The sabotage troops they were carrying were put to battle and wiped out to the last man.

Late in November, and behind locked doors, we burned all the papers, documents and even Norwegian money in the boiler in the Headquarters' office. For nearly three years the operation remained a closed book: we knew nothing of the fate of our friends or of the horrors which were perpetrated on the other side of the North Sea.

There was some bitterness and a feeling that the force had been sent on a suicidal mission. But complete secrecy prevailed.

While limited information was made available to SOE, 1 Airborne Division's last news until after the end of the war was the German press release that all had been killed in battle. Yet hope remained that there might be one or two survivors. A one hundred per cent death rate is not the norm in most battles. Were there neither wounded nor prisoners?

On 8th May 1945 the 400,000 strong German force in Norway surrendered. On 9 May, 1 Airborne Division was flown in, its 4,000 strength being dispersed over key cities. Three days after our landing, Lieut Jacobsen of the Norwegian Army and I set off from Oslo for Rjukan, to see for ourselves what had been involved in the operation and to see if there were any trails which might be followed. We were more fortunate than we deserved for, stopping at Kongsberg to discuss a route with the local Norwegian garrison, we stumbled across a member of the SOE Norwegian party, Knut Haugland, responsible for the glider reception. He explained that he had already reported fully to the War Office in London on his limited knowledge of events.

During that fateful evening the weather had deteriorated steadily and there was cloud cover with some snow. An hour or so before midnight he had heard a heavy aircraft overhead and he waited to signal to the descending glider- but none came. He had heard that the gliders had landed somewhere to the South-West. He knew nothing of the fate of the soldiers involved.

Haugland was most helpful and arrangements were made to visit Rjukan and the power plant three days later. A few of us, under the leadership of Major John Chivers, OC 261 Company went to both Rjukan and Vemork. Little that would help us discover the fate of the troops was found on this visit but a study of the location of the plant and the proposed approach from the lake by troops, unskilled in snow-warfare and survival made us wonder how and why such an operation had been conceived. (It must be remembered that this was May 1945 and the everyday use of the words 'atom bomb' was still in the future).

A second clue, which was to prove an important lead, came from a sapper travelling by train from Stavanger to Kristiansand. A casual conversation with a local railway porter led to western Telemark and to the discovery of graves, of firing squads and of bestial murder. 9 Company led the sapper search.

A glider released from the surviving aircraft crash-landed at Fylgjesdalen on a snow-covered mountain top. The officer in charge, David Methven G.M, his lance-serjeant, the two glider pilots and four others were killed immediately.

Four others were severely injured in the crash: they were taken by the German army to Stavanger hospital and then handed over to the Gestapo.

Five men survived the crash: they were tending their wounded when they were taken prisoner and removed to Grini concentration camp near Oslo.

The second glider, under command of Lieut A.C. Allen, and its towing plane crashed soon after crossing the Norwegian coast in the mountains north-east of Helleland. All of the crew of the plane were killed on impact. Only three of the glider force were killed in the crash. A number were severely injured and the Force was hopelessly adrift from its target. Some soldiers enquired of the Norwegians who found them the practicality of escaping, firstly to Sweden and then, when that was out of the question, the way to the North Sea coast. Whoever remained in charge decided to protect the injured, sent for a civilian doctor and surrendered. They were taken to the German camp at Slettebo near Egersund.

As the jigsaw of events was put together the story became more and more horrific. Eventually, over July and August 1945, all of the bodies discovered were reburied.

A black granite headstone was designed and erected in the cemetery at Stavanger. A carving on the stone of the winged Pegasus was followed by words for all of the Royal Engineers and the Glider Pilots who flew them.

> In Memory of Thirty Royal Engineers of the 1st British Airborne Divison, Two Glider Pilots of the Army Air Corps and Two Pilots of the Royal Australian Air Force, the occupants of two gliders which crashed at Helleland and Fylgjesdalen on 20th November 1942, whilst engaged on a gallant mission against the German atomic bomb research at Rjukan.
>
> *Og det er det stora*
> *Og det er det glora*
> *At Merket det stend*
> *Um Mannen hat stupa*

Freely translated, the verse of the Norwegian poet, Per Silvie, means: 'For this is the greatness and this is the glory, that the cause goes forward, even although the man falls.'

When the re-burials took place, we were still unaware of the full horror that surrounded the deaths of our fellow-sappers. Investigations by War Crimes teams completed the story.

Part II. The Aftermath

It was not just the legality of Operation *Freshman* itself that was questionable: so also was the treatment of the captured men. The chief of all military land, sea and air forces in Norway, General Nicolas Von Falkenhorst, together with other senior generals, had received from Adolf Hitler a personal instruction dated 18 October 1942, only a month prior to the *Freshman* operation.

In principle it was a direction that all raiding and sabotage troops, even if in uniform, were to be exterminated to the last man. If they surrendered, they were still to be executed. Leading German *Wermacht* generals such as Von Runstedt and Rommel ignored it.

It was essentially a 'no-quarter' directive: of the four critical paragraphs two were as follows:

> From now on, all opponents brought to battle by German troops in so-called Commando operations in Europe or in Africa, even when it is outwardly a matter of soldiers in uniform or demolition parties with or without weapons, are

> to be exterminated to the last man in battle or while in flight. In these cases it is immaterial whether they are landed for their operations by ship or aeroplane or descend by parachute. Even should these individuals, on their being discovered, make as if to surrender, all quarter is to be denied them on principle. A detailed report is to be sent to the OKW on each separate case for publication in the Wehrmacht communiqe.
>
> If individual members of such Commandos working as agents, saboteurs etc. fall into the hands of the Wermacht by other means- e.g. through the police in any of the countries occupied by us- they are to be handed over to the SDF immediately. It is strictly forbidden to hold them in military custody, e.g. in PoW camps etc., even as a temporary measure.

Of the further two, one excluded from the dictat all soldiers captured in open battle, all sailors captured after naval operations and airmen descending from a plane involved in an air-battle. The other threatened the Court Martial of any commander failing to implement the order.

It was almost a legal decree – but not quite. To be acceptable, such a command would have had to be made known not only to troops under command but also to the enemy. The illegality of the *Befehl* hinges on the fact that it was issued under 'Top Secret' cover. For implementation within his Norwegian command, von Falkenhorst had penned a savage addendum: 'If a man is saved for interrogation he must not survive his comrades for more than twenty-four hours.'

The British High Command was not aware of this directive at the time that Operation *Freshman* was launched.

The four wounded men put in the hands of the Gestapo at Stavanger suffered a horrific fate. All four were given a number of injections of experimental poison: three were then brutally strangled: the fourth was shot in the back. Their weighted bodies were dumped in the sea and never recovered.

The five captured men sent to the Grini concentration camp just outside Oslo lived only a few weeks longer. At dawn, on 18 January 1943 they were executed by firing squad.

The survivors and their wounded from the second glider taken to Slettebo were the first to die. Within ten hours of arrival all, including the severely wounded, were executed by firing squad. The wounded were first of all left by the roadside near the camp while the others were executed, one by one. Then the wounded were dragged to the site of the execution and shot. The bodies were thrown into a trench dug in the sand-dunes near Ogna.

After the end of the war the Germans involved were tried before War Tribunals. Von Falkenhorst was sentenced to death but the sentence was later commuted to twenty years imprisonment. Some of those who carried out the murders were hanged and some were shot: others were given long sentences of imprisonment. When the case of the individual members of the firing squad came to trial, Mark Henniker, who had albeit reluctantly mounted the raid, volunteered evidence in their defence. He submitted that if he, an officer with fifteen years of service, had the instinctive view that the operation was illegal, how was it that a group of young and inexperienced soldiers, probably conscripts with one or two years service, could be blamed for obeying their orders to shoot? It was a chivalrous gesture and typical of the man who made it, but the Court would have none of it. The President of the Court was a distinguished soldier, Major-General Coxwell-Rogers, who earlier in the war had been CRE 4 Infantry Division. The young members of the firing squad were each sentenced to thirteen years imprisonment.

Death in battle is an acceptable hazard of war: cold-blooded murder is not. The circumstances of these sapper deaths at the hands of the German Gestapo will always remain unforgivable. When Mark Henniker and I met on a number of occasions after the war, the subject of Operation *Freshman* and its justification or non-justification always came to the surface sooner or later. He was particularly concerned that he had been required to order a high-risk venture and then refused permission to command it. I know that it troubled his conscience, even as late as the end of the 1980s, only a few months before he died.

In the late-1970s Norsk Hydro, the plant owners and for whom I worked as a consultant, demolished the whole of the Vemork complex.

In later years I flew over the Rjukan site in a light plane with directors from that company. The pilot flew uncomfortably low and followed roughly the approach through the mountains which the gliders would have had to take. It was a difficult route but a skilled glider-pilot might have achieved it. On another day, in cool Spring weather, we walked the planned route which the sappers, untrained in mountain skiing or snow warfare, would have taken in conditions of heavy snow and ice. We studied the immediate vicinity of the plant and the proposed direction of the attack that the sappers were to mount into the plant itself.

It may not have been a 'mission impossible' but it was not far removed.

Report 20/7/45 Subject Operation *Freshman*
Maj PARDON (2) CA/G5 Division, Stavanger Zone
Ref my TPM RE 269 of 131500 B, copy to you

The only paper available from War Diary is the report on the operation submitted by Lt. Col MCA HENNIKER, DSO OBE MC RE then CRE, 1st Airborne Division, and I enclose extracts which may prove of use. The bulk of the report consisted of training, planning, and security arrangements.

Para 11. The Operation
The operation did not go according to plan. The weather forecast was reasonable enough, though not ideal, and events showed it to be accurate.

Aircraft 'A@ in which Gp/Capt COOPER DFC, RAF flew with S/ldr WILKINSON as Captain, left SKITTEN with a glider in tow at 1750 hours 19 Nov 42. The glider pilots were Sjt MPO STRATHDEE AAC, and Sjt P DOIG, AAC, and the half-section RE under the command of Lt METHVEN GM RE. On approaching the Norwegian coast, the REBECCA in the aircraft went unserviceable. The failure of the raid is due to this.

A landfall was made, and the aircraft flew towards the target, but could not find it. A second fix was made and a second attempt to find the target was made. There was barely enough petrol to get the tug and glider home.

The tug was able to climb above a cloud, and had to fly into it over an area about 40 miles NW of the target. Ice formed, both tug and glider lost height and at last the tow rope broke.

The wireless operator sent out a signal in clear at 2355 hrs - "GLIDER RELEASED IN SEA". This was picked up in WICK and at another RAF station, and an intersection of the bearings showed the position as over the mountains in SW NORWAY. At 0151 hours on 20 November the tug returned, and a careful check of courses and times proved that the glider was in fact released over land.

Aircraft 'E', captained by F/Lt PARKINSON RCAF, took off from SKITTEN with a glider in tow at 1810 hours 19 Nov. The glider pilots were P/O DAVIES RAAF, and Sjt FRASER RAAF. The RE half-section in the glider was under the command of Lt A ALLEN RE. Nothing further was heard of this, till at 2341 hrs a signal was received, believed to have been from the tug, asking for a course to bring it back to base.

The position at that time was found by intersection to be Lat 58 degrees 16 mins N 00 degrees 17 mins W. Nothing more was heard, and from first light on 20 Nov, two squadrons of Coastal Command aircraft and a destroyer searched the area all day without finding any trace.

On 21 Nov the following announcement was made over the German wireless, and repeated in the English papers-

'On the night 19/20 Nov, two British bombers, each towing a glider, penetrated SOUTHERN NORWAY. One bomber and both gliders were forced to land. The sabotage troops in the gliders were engaged and killed to the last man.'

From this, all that is certain is that the two gliders and the missing tug landed in NORWAY. It is impossible to have a battle in which there are no wounded, and it is possible- indeed probable- that a number of the men concerned are at large in NORWAY. The agents reported that no glider landed in the proper place.

PARA 14. Nominal Rolls and Next of Kin
The nominal rolls and next of kin of Army personnel are given in Appendix 'C' to the Operation Orders, with the exception of:-

320272 Sjt MPC STRATHDEE AAC
Next of kin: mother, Mrs AD Strathdee, 75, Grange Park Rd, Thornton Heath, Surrey.

3250420 Sjt DOIG AAC
Next of kin: mother, Mrs A Doig, 1172 Argyle Street, Glasgow
It is not proposed to notify next of kin until definite news is received.

Application for copies of Appendix 'C' to the Operation Order has been made to War Office by HQ 1st Airborne Division

(Haven't a bloody clue what that signature is… work it out yourself, good luck!)
Lt. Col, RE, CRE, 1st Airborne Division.

From War Office, 11th September 1945, to Mrs L B Bray
Madam,

With reference to your letter of the 6th August 1945, addressed to the Officer in Charge, Royal Engineer Records, Brighton, I am directed to state that your husband, No. 1884418 Lance-Corporal P.W. Bray, Royal Engineers, was a member of a party of men who took off from this country in gliders on the evening of 19th November 1942, for a special operation over Norway. Unfortunately, the weather conditions over the target were bad and according to reliable information received (*hand-written inserted*, we affirmed that) the glider in which your husband was travelling crashed into the sea off the coast of Norway near Egersund.

I am to convey to you an expression of the Department's sincere sympathy in your great loss.

I am, Madam,
Your obedient servant. (V Hawley?)

3. *'Leif Tronstad – the Heavy Water Hero'*
Source: *Chemistry World*, 7 February 2018, Professor Lars Öhrström, Chalmers University of Technology

Leif Tronstad was once called the worst student ever accepted into the chemical engineering programme at the Norwegian University of Science and Technology (NTNU) in Trondheim. He turned out to be perhaps its most brilliant graduate. By the late 1920s, the young inorganic chemist used ellipsometry to show some metals had a thin, protective oxide layer. Today, it's a standard technique in thin film chemistry and physics – at the time, it solved why some metals, contrary to their electrode potentials, will not corrode. But Tronstad's name will forever be attached to something else: heavy water.

After Harold Urey discovered deuterium in 1931, Tronstad quickly realized that there would be both a use, and a market, for this exotic atom in several branches of science. But obtaining it was a tricky problem: only 0.01% of hydrogen atoms are deuterium. Pure samples were acquired through the production of heavy water, D O, by water electrolysis: the slightly faster formation of H than HD and D concentrated heavy water in the sample, but initially the electricity cost was forbidding. Then Tronstad and Jomar Brun, the head of Norsk Hydro's hydrogen electrolysis factory, came up with a commercially viable way of producing heavy water by using cascade electrolysis, powered by inexpensive Norwegian hydroelectric power.

The duo arranged to set up a heavy water installation at the plant in Rjukan, close to Vemork in Telemark. It was one of Tronstad's many interests in the Norwegian chemical industry and might have been unremarkable if it had not been for the second world war. In 1940, Germany invaded Norway. Tronstad, a dedicated patriot and officer in the Norwegian army reserves, joined the brief defence. As the Germans occupied his country and the armed forces surrendered, Tronstad took part in the *Hjemmefronten* (Home Front) resistance, organising students and staff at NTNU. He was soon living dangerously. In September 1941, he narrowly escaped the Gestapo and fled to London, leaving his wife and two children behind. They corresponded throughout the war via various clandestine operations, but they would never meet again.

Tronstad kept abreast of military-related science developments, liaised with MI6 and even had time to go out nightclubbing with Vivien Leigh. The Norwegian government-in-exile and royal family had established themselves in London. Here, Tronstad found himself in the middle of enthusiastic patriotism, but also in a complex web of mistrust where everyone could have ulterior motives. The Norwegian cabinet did not trust the resistance; the resistance didn't trust the Norwegian army; the Norwegian units in the Allied Special Operations Executive (SOE) didn't trust their British officers. Tronstad was not above motives of his own.

Nazi Germany was aiming to develop its own atomic bomb and needed D O from Rjukan to slow down neutrons and sustain a nuclear chain reaction. The Allies couldn't let this happen and so planned to bomb the plant Tronstad had established, but he opposed the move, suspecting it would hand over a world monopoly of D O to Imperial Chemical Industries after the war. As Tronstad said: 'Blood is thicker even than heavy water!'

In the meantime, Tronstad's diary (kept secret throughout the war) records a chemist with an extraordinary life. He talked with ministers, met with the Norwegian king and crown prince, took night trains up to Scotland to sort things out between the Norwegian SOE troops and their officers, and undertook Commando training. He quickly advanced in rank, ultimately reaching major. Eventually, Tronstad agreed to blow up his brainchild.

In February 1943, SOE launched Operation *Gunnerside*, destroying the plant in arguably the most successful Allied sabotage action of the war. Tronstad was not allowed to participate but took part in the planning (as did Brun). The story has been told many times – perhaps most famously in *The Heroes of Telemark*, where Kirk Douglas plays a fictional amalgam of Tronstad and the leader of the *Gunnerside* team, Joachim Rønneberg (although, for unfathomable reasons, his character was a physicist).

Tronstad and fellow SOE officer John Skinner Wilson both received OBEs for the planning of *Gunnerside*, but Tronstad ached to get into action himself. One reason was probably a desire to share the dangers and hardships of the men and women of the Allied resistance movement, many of whom he had himself sent into action, and many of whom were lost in action or killed at the hands of collaborators or the Gestapo. In October 1944, he finally got his chance as part of Operation *Sunshine*. He was parachuted onto the Hardangervidda mountain plateau, where important industry and infrastructure in southern Norway had to be protected against sabotage by retreating German troops.

It was right up Tronstad's alley – his interest in industry and still existing professional network made him the ideal commander. But disaster struck. ... A month before the German surrender in Norway, Tronstad was questioning a Norwegian collaborator in a remote hut when the man's armed brother turned up. In the ensuing firefight Tronstad and fellow SOE officer Gunnar Syverstad were killed. Tronstad was 41. He was probably 'The greatest post-war leader Norway didn't have.'

4. *'Operations Gunnerside and Grouse – Special Operations during World War II against the German Controlled Heavy Water Plant in Norway'*

Author: Major Frode Kristoffersen, Norwegian Army, United States Marine Corps Command and Staff College

> *Thesis*: The operations were a success due to their execution by Special Operations Forces, which demonstrated that a small, select group of individuals with specialized training and thorough preparation could succeed where a conventional force could not.

Discussion: The production of heavy water at the German controlled plant at Vemork, Norway was· essential for Hitler's nuclear research and development of a nuclear capability during World War II. German plans for tripling the production in 1942 made the Allies fear that Hitler was about to win the race for the atomic bomb. Churchill prioritized the destruction of the heavy water plant. In 1942, British plans for a Commando raid at the heavy water plant included the reconnaissance party, Operation *Grouse*, formed by four Norwegians from Special Operations Executive (SOE) and a glider-inserted strike force, Operation *Freshman*, formed by combat engineers from 1st Airborne Division. After the tragic failure of Operation *Freshman* on 19 November 1942, SOE decided to leave Operation *Grouse* in place in order to function as a combined reconnaissance and advance party for a Special Operation; Operation *Gunnerside*.

After a total of four winter months of reconnaissance, waiting, and surviving in the mountains of Norway, Operation *Grouse* could finally link up with Operation *Gunnerside* on 23 February 1943. Together, these two patrol-sized missions successfully destroyed the heavy water production in February 1943, and were able to withdraw without firing a single shot. The teams overcame the frictions represented by German forces and the harsh winter of 1942 to 1943 firstly as a result of mastering the operating environment. The teams formed by a total often Norwegians had all been through extensive selection and training and possessed excellent winter skills in addition to local knowledge. Secondly, preparations for the specific objective went on for months.

These preparations enabled the two small teams of Operations *Gunnerside* and *Grouse* to infiltrate uncompromised into the objective and execute a surgical demolition of the most vulnerable part of the heavy water production. Furthermore, the thorough preparations ensured flexibility when the teams had to adjust their plan and resulted in flawless execution. The third main reason for the success of Operations *Gunnerside* and *Grouse* is the ten soldiers' strong motivation for their mission. The soldiers faced high risks during the strike and Operation *Grouse* overcame extreme hardships during their prolonged mission in the snowy mountains.

Conclusion: Operations *Grouse* and *Gunnerside* achieved an outcome of strategic importance when they destroyed the heavy water production at Rjukan in February 1943, in an environment that proved too challenging for a conventional approach. The small group of specially selected soldiers capitalized on the soldiers' specialized training, skills, and preparations as well as their unique motivation for solving the mission in their occupied homeland.

During my studies of the amazing operations against the heavy water plant at Vemork in 1942 to 1943, I received the sad news that two of the participants had died, of natural causes. Knut Haugland died on Christmas day 2009 at the age of 92. It was the end of a remarkable life, including the Kon Tiki expedition where he crossed the Pacific in a primitive balsa boat as an archaeological and social experiment. Jens Anton Poulsson, a retired colonel and regimental commander, died on 2 February 2010. Both had lived long and remarkable lives, and I am especially excited about the story of their 4-month long reconnaissance mission before the strike at Vemork. Furthermore, the more I study the Vemork operations, the more I realize the striking similarities with current theories of special operations.

I have studied two modem theories of Special Operations during my research. The authors of these theories did not study the operations at Vemork when developing their theories. Nevertheless, the two team leaders, Poulsson and Joachim Ronneberg of Special Operations Executive, executed operations that comply with current theories. Operations *Grouse* and *Gunnerside* were textbook Special Operations carried out in challenging terrain and weather, despite the presence of an overwhelming German force. I had the honour to interview Poulsson during my studies and I also received input from Ronneberg. These inputs from primary sources have proven immensely valuable for my work, and I am forever thankful to them both for so willingly sharing their experiences.

Introduction: The small town of Rjukan, Norway and its fertilizer plant at Vemork, received the unlikely attention of Hitler, Churchill, and Roosevelt during World War II because of the race to develop the atomic bomb. German nuclear research relied on heavy water[1] and after the invasion of Norway in 1940 Hitler controlled Vemork, "the world's only source of large-scale production [of heavy water.]"[2] Thus, Vemork became a target of strategic importance. Although scientists in the United States did not use heavy water as part of their nuclear research, the Allies' assumption was that it was necessary to shut down the plant in order to halt German progress in their nuclear program.

In 1942, British plans for a Commando raid at the heavy water plant included the reconnaissance party, Operation *GROUSE*, formed by four Norwegians from the British Special Operations Executive (SOE), and Operation *FRESHMAN*, a glider inserted strike force from the British 1st Airborne Division. After the tragic failure of Operation *FRESHMAN* on 19 November 1942, SOE decided to leave Operation *GROUSE* in place for further reconnaissance in order to set conditions for a small-scale special operation, Operation *GUNNERSIDE*, conducted exclusively by Norwegians from Kompani Linge of SOE. Operations *GROUSE* and *GUNNERSIDE* were patrol-sized missions that together led to the successful destruction of heavy water production in February 1943.

After four months of reconnaissance, Team *GROUSE* linked up with the parachute inserted Team *GUNNERSIDE* and conducted the strike at Vemork without firing a single shot. The small but highly motivated and skilled groups were able to overcome the frictions the operation faced and succeed in the strategically important mission of 1 destroying the heavy water production plant. Operations *GUNNERSIDE* and *GROUSE* were a success due to their execution by Special Operations Forces (SOF) which demonstrated that a small, select group of individuals with specialized training and thorough preparation could succeed where a conventional force could not.

An examination of Operations *GUNNERSIDE* and *GROUSE* demonstrates fundamental and timeless qualities of special operations in which a small unit can achieve strategic effect: furthermore, the two operations are highly relevant for current operations by providing guidance for SOF on the importance of quality in training, selection, and preparations. Operations *GUNNERSIDE* and *GROUSE* illustrate Robert Spulak's theory stating that the strength of SOF is primarily derived from the ability to "address the ultimate sources of friction."[3] Today, SOF units need to master challenging operating environments in order to maintain their qualities as creative and flexible forces, to remain an alternative ·to conventional forces for missions of strategic importance similar to Operations *GUNNERSIDE* and *GROUSE*.

Current theories of special operations Looking at Operations *GUNNERSIDE* and *GROUSE* through the special operations lens, two different views may explain the successful outcome of the operations. Admiral William H. McRaven's SPEC OPS, assesses principles that lead to success for a small force in an offensive operation, whereas Dr. Robert G. Spulak's essay "A Theory of Special Operations" focuses on the qualities of the personnel that are needed in order to succeed. These two positions complement each other and explain how Operations *GUNNERSIDE* and *GROUSE* achieved success.

McRaven's six principles for special operations are: surprise, speed, purpose, security, repetition, and simplicity.[4] From eight special operations case studies, McRaven derived that the key to success is "a simple plan, carefully concealed, repeatedly and realistically rehearsed, and executed with surprise, speed, and purpose".[5] By following these principles, SOF personnel are able to overcome the frictions of war if they also hold the "moral factors of courage, intellect, boldness, and perseverance".[6] Spulak's theory is focused, as well, on overcoming friction in special operations but he contends that the attributes of SOF personnel are the key to success rather than principles of execution. Spulak focuses on attributes needed to overcome Clausewitzian friction as the key to achieve success.[7] In *On War* Clausewitz defines friction to be "the force that makes the apparently easy so difficult."[8]

This paper uses Clausewitz' definition of friction and puts special emphasis on his further description of friction: the "countless minor incidents- the ones that you can never foresee- [that] combine to lower tlre general level of performance."9 In Spulak's view, tlre three vital attributes, or qualities, of SOF personnel are: "elite warriors, creativity, and flexibility."10 Through extensive selection and training, along with prioritizing quality over quantity, SOF units may create teams witlr personnel who possess these attributes to a larger extent than personnel in regular units.

As a result, SOF teams may overcome friction to a larger degree than a conventional unit, which gives the SOF team a broader span of operational capabilities and fewer limitations. Together, McRaven's and Spulak's theories demonstrate how a small SOF team of specially selected and trained personnel may achieve success. Where a conventional unit may be limited by the specific challenges of an operation, SOF units witlı special 3 skill sets may overcome these challenges and use them to their advantage (e.g., covert parachute insertion of a small team where the terrain limits movement of larger conventional forces).

Moreover, the limited size of a SOF unit may also benefit the operation, since smaller teams have a better chance to avoid enemy surfaces, operate covertly, and achieve the element of surprise. As a result, a successful SOF team may bring to bear a superior relative strength by engaging the unprepared enemy with the advantage of initiative and precision targeting. At Vemork, Operation *FRESHMAN* experienced severe challenges due to the terrain and weather in Norway. Conversely, the SOF operations depended on the terrain and weather to get to the objective without being compromised by the numerically superior Gelman guard force. It was especially the SOF characteristic of certain access that proved to be vital for the success of SOE' s operations at Vemork.

As explained by Spuıak, certain access is a SOF characteristic that enables SOF to operate in areas where conventional forces may not operate.[11] Thus, Operations *GUNNERSIDE* and *GROUSE* provide an excellent opportunity to validate McRaven's and Spulak's views. The strike at Vemork depended on adherence to all McRaven's principles. Moreover, the members of the operation possessed the qualities needed for SOF personnel. Britain's SOE limited the total number of its operatives through selection, training, and preparations to ensure that they held those qualities listed by Spulak. The Heavy Water Plant at Vemork Before World War II Germany was a leading nation in nuclear research. Hitler needed large quantities of heavy water for his nuclear program, since German scientists used it for slowing down nuclear reactions during testing. Norsk Hydro led the world in 4 the production of heavy water by 1940, as a by-product of its fertilizer production at. Vemork. After- the invasion of Norway on 9 April1940, German forces controlled Vemork and heavy water production.

German plans for tripling heavy water production in 1942 led the Allies to conclude that Hitler was close to winning the race to create an atomic bomb. Based on the assessment of Hitler's progress, the British War Cabinet, approached Combined Operations Command with an urgent request that Vemork be attacked with sufficient force to destroy all accumulated stocks of heavy water, the major pieces of machinery in the electrolysis plant used for its production, and the power station situated at the rear of the plant.[12]

Combined Operations Command examined six different courses of action for the mission.[13] All six were considered high risk for the force involved except for the option of bombing the plant; nevertheless, plans for bombing Vemork were soon abandoned. The plant situated in a steep and narrow valley; Combined Operations Command concluded that there was a low probability of hitting the target. As the most important part of the production facility was underground even a precise hit would mean only a short delay in heavy water production. Moreover, the risk of collateral damage if stray bombs hit the populated area of Rjukan and its highly explosive liquid-ammonia storage tanks mitigated against bombing.[14]

... Under pressure from the War Cabinet to find a solution, Combined Operations Command developed Operation *FRESHMAN* to destroy the heavy water production plant at Vemork. The plan was to insert two groups of Army engineers from 1st Airborne Division by gliders. The glider insertion would allow the groups to avoid the concentration of German soldiers in the town of Rjukan and attack the small guard force at Vemork directly. The two groups were to demolish the underground processing plant and, if possible, also the electrical plant.[15] Combined Operations Command chose engineers for the operation because of the importance of the demolition job. Only well-placed charges would do substantial damage to the robustly constructed heavy water plant. The two 12-man teams constituting Operation *FRESHMAN* were all volunteers who had little or no knowledge about the importance of the operation.

When British Combined Operations Command saw the need for a reconnaissance party for Operation *FRESHMAN*, it turned to SOE to employ Norwegian operatives with local knowledge and winter skills. The reconnaissance part, codenamed Operation *GROUSE*, was given the mission to reconnoitre and prepare a landing strip, receive the gliders, and guide the strike force to the objective. SOE appointed a four-man team of Norwegians, all born in Rjukan, for the mission. On 17 October 1942, Operation *GROUSE* parachuted into the snowy mountains of Norway, but landed miles away from the planned drop zone. This put them a considerable distance from the objective. Despite heavy challenges due to the faulty parachute drop, communications issues and bad weather, the Norwegian SOE operatives were ready to receive the British engineers during a favourable moon phase one month later.

On 19 November 1942, the *GROUSE* team's month of preparations proved to be in vain when Operation *FRESHMAN* failed. Weather conditions during the night of the insertion led to fog and icing on the two gliders and the Halifax planes towing them. One plane crashed still towing the glider and the other plane was forced to release the glider in the blind before returning to England. All three crews from the two gliders and the crashed plane were killed. The majority died during the crash landing; nevertheless, German soldiers captured and killed the few survivors. The execution of these British soldiers was the main reason for the British condemnation of General Falkenhorst, the German commander-in-chief in Norway, as a war criminal after World War II.[16]

For Operation *GROUSE*, the tragic end of the forty-one British soldiers was not known but the team assumed that Operation *FRESHMAN* failed. Despite the loss of the engineer contingent, SOE decided to maintain the *GROUSE* team as a reconnaissance and advance party for a new operation, Operation *GUNNERSIDE*.[17] The new operation would be carried out solely by SOE and Kompani Linge, as SOE believed "there was a chance of success using clandestine methods."[18]

Operation *GUNNERSIDE* depended upon maintaining the *GROUSE* team in country and the four-man patrol stayed put. They lived off the land for the next three months and experienced several delays when bad weather during the favourable moon phases did not allow for the parachute-insertion of Operation *GUNNERSIDE*.[19] Finally, on 23 February 1943, the two teams linked up. The combined group numbered a total of ten Norwegians predominantly from the local area. After coordination and consolidation of the final plans, the ten-man team carried out the strike on the night of 28 February.[20]

By infiltrating via steep terrain and armed with knowledge of the German guard routines, the team entered the heavy water plant and placed explosive charges without being compromised. The team executed the operation successfully without firing a shot and exfiltrated the Vemork area. From their mountain operating base, the team started their extraction by skis to Sweden, a journey of approximately 400 kilometres[21] that they made in fourteen days.

From neutral Sweden, the group returned to England via airplane. While Operation *FRESHMAN* failed during insertion and proved the military challenges of attacking Vemork, the success of SOE's operations demonstrated that SOP was best suited for achieving Allied objectives at this strategically important target located in "some of the most dangerous and inhospitable terrain known to man"[22] and heavily occupied by German troops.

Selection and Training Winston Churchill established SOE in 1940 in order to train resistance movements and carry out operations in German-occupied countries.

Norwegians serving in SOE were organized in Norwegian Independent Company Number 1, later named Kompani Linge, after the company commander. Captain Martin Linge was killed in the December 1941 Måløy Raid. British forces had to start from scratch when establishing SOE and experienced several setbacks while building the unit under fire.

Despite such an inauspicious start for SOE, by 1942, when Colonel John. S. Wilson was appointed the head of the Norwegian section, "its great potentialities were becoming clear"[23] and the unit became a true SOP unit even when evaluated by twenty-first century criteria.[24] According to Spulak it is important that SOP units emphasize selection and training of personnel with the right qualities. SOE sent their Norwegian recruits through a series of selection and training programs before the most talented made it to the operations branch.

Undesirable recruits were sent to other military units, and, as the British author Ray Mears put it in his description of the selection process, "it was surprising, given the desperate need for recruits, how ruthless SOE were in culling the numbers."[25] SOE understood that quality had to be prioritized in building a SOF unit. By thorough and varied selection and training programs, SOE officers ensured that the operations branch only consisted of operatives who were able to adapt and learn during various challenges. As Spulak would characterize SOF operators, SOE operatives held the qualities needed to overcome friction; they were creative and flexible elite warriors. Selection and training took place in England, but getting there was a test of determination and will in itself.

Due to the German occupation of Norway, getting to England was a challenge for most Norwegians who volunteered to fight the German occupation force. SOE recruited some Norwegian members in Sweden and were able to fly a limited number to England directly from Stockholm. Due to limitations in crossing the battle areas, the majority of Norwegian volunteers had to travel a long circuitous route to England.[26] For example, two of the *GROUSE* team members travelled from Norway to England through Bombay.[27]

Spulak recognizes "self-selection" as a possible way to select personnel with SOF attributes and suggests that "part of selection of attributes is self-selection in who is motivated to volunteer for SOF and who is motivated to stay."[28] Norwegians that made the troublesome journey to England in order to volunteer for fighting the German occupation force must have had a strong motivation for their cause. Travel to England was not a formal part of the selection, but it was a means of efficient self-selection. In the formal selection in England, SOE initially recruited personnel to its operations branch by interview. Norwegians who had distinguished themselves in the resistance against the German invasion or occupation or had done so otherwise were approached by recruitment officers. Second, soldiers were selected during training.

To become a part of the operations branch of SOE, soldiers had to pass four stages of training.[29] At all stages, the soldiers could be considered disqualified from further progression. Altogether the selection and training program ensured that the operations branch consisted of SOF personnel of high military standards.

Overcoming friction SOB managed to select and train teams with the right qualities to overcome the extreme friction presented by weather, terrain, and the enemy in the Rjukan area. When the environment of Vemork is considered, overcoming friction was a most challenging task for the SOE team. In January 1943 the German guard force at Vemork numbered thirty soldiers. After discovering Operation *FRESHMAN*, the Germans reinforced the defence of the plant with a minefield.[30] Furthermore, weather and snow conditions during winter, combined with steep terrain, left the Germans to believe that the road through Rjukan was the only likely avenue of approach to the heavy water plant.[31]

In Rjukan, the German forces numbered 20,031 and the Germans also controlled a network of sympathizers among the local Norwegian population. Jens Anton Poulsson, leader of Operation *GROUSE*, still believed that the conventional Operation *FRESHMAN* could have succeeded if only one of the two gliders had been able to land on the prepared landing site.[32] Nonetheless, both the plans and the outcome for Operation *FRESHMAN* show how much the operation was limited by the frictions of the operating environment.

The engineers lacked winter warfare skills, which made the plan marginally feasible and left no room for contingency plans. SOB's initial objections to the plans for Operation *FRESHMAN* demonstrate that the newly established SOF community already understood the importance of mastering the operating environment. All SOE objections were based on the impact of terrain and weather on the operation;[33] moreover, the conventional soldiers of the 1st Airborne Division had few contingency options if their primary plan failed. First, glider insertion limited flexibility once the group was over Norway. A return to England was possible in theory but fuel limitations and increased vulnerability from German anti-aircraft guns meant that the tow planes most likely would have to release the gliders once over Norway.

Thus, the gliders crossed the practical point of no return before entering Norway,[34] leaving Operation *FRESHMAN* without the flexibility of Operations *GUNNERSIDE* and *GROUSE*. For the SOE parachutists, the decision point was the jump, which meant that they had the option to abort the mission at any stage during the flight, in case of last-minute changes in the situation. Indeed, in January 1943 the team had to abort an insertion attempt over the drop zone because of deteriorating weather conditions.[35]

Second, the lack of winter warfare expertise in general and skiing skills in particular, left Operation *FRESHMAN* with limited flexibility to adjust the plan if the landing had been successful. Lack of winter warfare and survival skills meant that the British engineers may not have been able to extract even in the case of a successful strike. According to Poulsson, "*GROUSE* men always considered it likely that the *FRESHMAN* personnel were on a one-way ticket. They would never have managed to get across to Sweden."[36]

Moreover, 1st Airborne Division's engineer's limited ability to master the winter conditions led to the decision that if a soldier could not keep up with the unit, his fellow soldiers should give him morphine and leave him behind.[37] Altogether, Operation *FRESHMAN* was a bold venture that had a narrow chance of success. Combined 11 Operations Command's decision of launching two teams of engineers for the mission was a means for increasing the likelihood of a successful outcome, since a single team was deemed sufficient for the task. This decision was also telling of the views of two different communities: the British conventional forces priority on mass versus SOE's full priority on quality. In stark contrast to *FRESHMAN*'s limitations in the operating environment of winter Norway, stands *GROUSE*'s ability to adapt to changing plans, roles, and timelines during their four months of efforts in the same winter conditions.

Both *GUNNERSIDE* and *GROUSE* personnel were selected from the operations branch of SOE by their team leaders Ronne berg and Poulsson. These two officers had been functioning as instructors within SOE and had thorough knowledge of the Norwegian personnel. In addition to demolition skills, Ronneberg lists ski skills and cooperative skills as the most important criteria for selecting the members of the two teams, adding that "he had personal knowledge of each member and knew what they were good for."[38] Both SOE's appointment of skilled team leaders and these leaders' selection of team members were focused on finding the men that were best suited for the specific mission at Vemork.

Thus, SOE's personnel assigned to the operations were experts on the operating environment that was so unfamiliar to the *FRESHMAN* personnel. The terrain and weather were the main sources of friction in the Vemork raid, and the flexibility SOE gained by selecting soldiers that were highly trained for the operating environment proved to be of utmost importance for the operation. The priority SOE put on finding the right men for the environment at Vemork complies with Spulak's theory on SOF success by addressing the ultimate sources of friction and selecting appropriate attributes to overcome those sources.

SOE's priority on winter warfare experts shows that they assessed the critical phase of the operation different from Combined Operations Command. Whereas Combined Operations Command prioritized actions on the objective by forming Operation *FRESHMAN* of combat engineers, SOE prioritized the task of getting to the objective, Spulak's certain access. In hindsight, SOE clearly understood the military challenge at Vemork best. The operation called foremost for winter warfare experts, while the specific skill, the demolition on the objective, proved to be of secondary importance.

Even though the SOE members had fairly skilled demolition experts in the team, they were not engineers. However, the specific demolition of the heavy water apparatus could be exercised during preparations for the operation, and the technicalities of the demolition itself did not turn out to be a source of friction for the operation. Operations *GROUSE* and *GUNNERSIDE*'s ability to address the ultimate source of friction, the operating environment, provided the operation with the element of surprise and enabled the small force to avoid German guards on the objective as well as German troops in general in the surrounding areas. The group negotiated terrain during the bold infiltration to and exfiltration from the target area that was so steep that the German guards at Vemork had deemed the terrain impassable.[39]

Furthermore, the operation adhered to McRaven' s principle of speed. German forces were rendered ineffective as they discovered the operation only after the demolition was completed and the *GUNNERSIDE* team was withdrawing. In addition to the wise choice of selecting winter-skilled Norwegians – predominantly locals from the Rjukan area – to carry out Operations *GUNNERSIDE* and *GROUSE*, SOE's training program also proved important for the success of the mission. Kompani Linge's general training took place in Scotland, "where the terrain most closely approximated to the wilds of their homeland and offered the only place in Britain where they could practice cross-country skiing for significant periods of the year."[40]

Furthermore, the general training was focused on small unit operations, where self-sustainment was central. The teams trained for carrying all their own supplies, and all skill sets needed for the missions, including special demolitions, medical expertise, and communications, had to be covered by the members of the team. In order to be ready for covert insertion, SOE put all personnel through parachute training. In sum, the personnel from Kompani Linge were highly specialized for their task, small-unit operations in their homeland, Norway.

By focusing on environmental skills and self-sustainment, SOE provided training that produced Spulak's "elite warriors" for the Norwegian operating environment. These self-sustained teams of winter warfare experts were well prepared to overcome frictions to the operations. Together with the winter skills, command relations supported Operation *GROUSE*'s ability to operate independently, when mission-critical communications with SOE broke down. The communications equipment was key for Operation *GROUSE*, both for coordination of the pre-planned mission, and especially for receiving new plans and orders after Operation *FRESHMAN* failed.

Operation *GROUSE* had an experienced telegraph operator – according to SOE the best they bad in the unit during World War

II[41] – but the communications set had limitations, resulting in long periods without guidance from SOE, while the team struggled to get communications running. Team leader Poulsson's ability to lead *GROUSE* through these uncertainties showed his qualities as a leader and his thorough understanding of the mission's intent, but it also proved the importance of *GROUSE*'s winter expertise. The ability to fully master the operating environment provided the team with enough time to re-establish the mission-critical communications with SOE.

Motivation, McRaven lists a strong sense of purpose, or motivation, as one of the main factors needed for a special operation to succeed. Based on this motivation, "the moral factors of courage, intellect, boldness, and perseverance have to ... prevent the frictions of war from ... causing defeat."[42] Operation *GUNNERSIDE* was a bold venture where the participants knew the high risks involved in the face of numerous German soldiers. Even though the courage of the conventional soldiers volunteering for the risky Operation *FRESHMAN* was admirable, Operations *GUNNERSIDE* and *GROUSE* tested courage and perseverance of the Norwegian soldiers to extremes. A strong motivation was the foundation that helped the members of these SOE operations overcome the hardships they faced at Vemork.

The Norwegian members of SOE had already been through several situations testing their motivation. The only reason they fled to neutral Sweden after organized resistance in Norway ended was to proceed to England and to have the chance to continue the fight from there. Furthermore, they had all passed the hardships of training and selection needed to become a member of the operations branch of SOE, but Vemork was the ultimate challenge that exemplified the importance of motivation for participants in a special operation. 15 The members of *GROUSE* did not fully understand the use or importance of the heavy water at that time, but they understood the end state of the mission and were very determined to achieve it. 'This motivation was tested to the extremes and proven during Operation *GROUSE*'s four-month long reconnaissance mission.

The conditions Operation *GROUSE* faced were so harsh that the team members had physical signs of undernourishment when they linked up with Operation *GUNNERSIDE*. Discoloured skin from vitamin deficiency added to their skinny, ghostlike appearances at this stage. The group had pushed their physical limits so far during Operation *GROUSE* that after the operation, one of the members was diagnosed with beriberi, a serious illness resulting from malnutrition.[43] In these conditions, and after the failure of Operation *FRESHMAN* had rendered original orders for Operation *GROUSE* obsolete, no one could have blamed the team if they had aborted the mission and left the extreme winter conditions.[44] Instead, the team was relieved to receive the message from SOE to stay put for a new operation.[45]

Operations like *FRESHMAN* that gathered volunteers regardless of the limited knowledge the soldiers had of the risks involved displays courage based on a strong motivation to fight Hitler's forces. Such courage is seen throughout many Allied operations in World War II; however, Operation *GROUSE* remains special. The members displayed a degree of perseverance that few other operations can match, demonstrating a distinction in the type of motivation needed for special operations and conventional operations.

Since SOF normally operates in small, self-sustained groups that are highly dependent on all its members, every SOF member needs to possess a strong motivation for fulfilling the end state of the mission. Furthermore, the strategic importance of SOF missions and the high risks involved call for highly dedicated personnel. McRaven argues the importance of purpose for special operations whereas Spulak's theory further distinguishes between conventional and special operations in their motivation and risk. Spulak's statement that "special operations are missions to accomplish strategic objectives where the use of conventional forces would create unacceptable risks due to Clausewitzian friction"[46] illustrates not only that SOF missions accept high risks but also that all SOF missions should be of high importance. Since SOF personnel are expected to face and mitigate high risks, the soldiers will play an important role.

Consequently, SOF personnel's strong motivation is both a result of the importance of a special operation mission and an element that is needed in order to face high risks. Although a strong motivation fosters courage to execute missions and fulfil the commander's intent, SOF should aim beyond commander's intent. Because of the high importance of special operations, failure may be fatal. Hence, SOF should mitigate the risk of failure by flexibility. In order to become the flexible force that can handle "informational uncertainties and unforeseeable differences between perceived and actual reality"[47] SOF must expect that even the commander's intent[48] might change during a mission.

When calling for flexibility, Spulak writes that "we can't know what's out there."[49] The uncertainty of the battlefield means that creativity and flexibility are important means to handle major changes to the plan. For Operation *GROUSE*, creativity and flexibility was the difference between mission failure and success. The members' motivation for achieving the end state was stronger than merely a spark for igniting courage; it was a lasting quality that gave them endurance to push themselves to limits of physical damage, even when all initial plans appeared obsolete.

The benefit of Operation *GROUSE*'s flexibility and will to carry on with their mission was that it gave SOE time to plan for a new mission after Operation *FRESHMAN*'s failure. In an interview on 10 June 2009, the team leader explained how he was able to withstand the physical and mental strains during four months of reconnaissance prior to the raid on the objective, stating that "my motivation was love for King and country."[50] Colonel Poulsson's answer shows how determined he was in destroying the heavy water production. Motivation to withstand the challenges Operation *GROUSE* faced during their four-month stay in the harsh winter of 1942 to 1943 goes beyond military professionalism. It was also an extreme motivation for achieving the end state of the mission.

Preparations for Operations *GUNNERSIDE* and *GROUSE* were thorough not only for specifics of the operations like the demolition but also with regards to more general matters like winter equipment. Specific preparations towards the Vemork objective were "indispensable in eliminating the barriers to success", as emphasized by McRaven. Furthermore, general preparations enabled the operations to maintain their flexibility, one of the three key SOF qualities of Spulak's theory. The key elements of the preparation for the strike at Vemork were the reconnaissance mission and Operation *GUNNERSIDE*'s rehearsals. Looking at the work put into preparation before the strike at Vemork, it is evident that SOE understood the utmost importance of detailed rehearsals for a special operation. Preparations followed both McRaven's principle of repetition and Spulak's emphasis on flexibility made possible by discovering the "ground truth."[51] In Operations *GROUSE* and *GUNNERSIDE* erroneous parachute drops challenged the plan from the moment the personnel landed but thorough preparations enabled the groups to carry on with their missions. This ability to plan for contingencies gave the SOF missions a flexibility that the conventional operation, Operation *FRESHMAN*, lacked.

Operation *GROUSE*'s four-month long reconnaissance mission ensured valid and timely information for the planning and execution of Operation *GUNNERSIDE*. The extensive reconnaissance period was a result of circumstances that could not have been foreseen: the tragic failure of Operation *FRESHMAN* and the several weather delays of Operation *GUNNERSIDE*. Consequently, most of the time at the mountain plateau was a matter of waiting and survival; nevertheless, the reconnaissance mission was vital for the strike at Vemork.

In addition to providing information on the situation in the target area that supported Operation *GUNNERSIDE*'s preparations in England, Operation *GROUSE*'s contribution to the operation after link up was substantial. The team leader, Poulsson and his men harboured the *GUNNERSIDE* team into a safe cabin from where they would carry out the last coordination of the plans. The reconnaissance party's views were incorporated, and the operations team adjusted its infiltration and exfiltration plans accordingly.

In a 2009 interview, Poulsson observed with characteristic understatement in his response to the question of whether the operation could have succeeded without the reconnaissance that "it would have been hard."[52] Operation *GUNNERSIDE* did not waste time while waiting for insertion. Rehearsals during their preparations in England were carried out on a "mock-up model of the apparatus ... which was an exact exterior copy of the real target."[53] According to McRaven, a full-dress rehearsal like the training Operation *GUNNERSIDE* went through is "essential to the success on the battlefield"[54] and this proved to be true for Operation *GUNNERSIDE*.

Furthermore, as McRaven states, "repetition, by its very nature, improves speed on the target."[55] This speed was most likely the reason why the small group was able to place the charges and withdraw before the German guards could react. SOE used a Norwegian scientist, Professor Leif Tronstad, who had detailed knowledge about Vemork and Germany's use of heavy water, as an advisor during the training for Operation *GUNNERSIDE*.[56] The thorough training that Operation *GUNNERSIDE* undertook as a result of this knowledge about the objective is in accordance with the well-known quality of precision in special operations.[57]

This knowledge enabled the small group to capitalize on precision in lieu of mass and infiltrate the important high concentration installation uncompromised. The training on a full-scale model also enabled the *GUNNERSIDE* team to place their charges only at the most critical and vulnerable structures of the heavy water production plant: a surgical pinprick with strategic effects. Moreover, the team's detailed preparations were keys to success in the contingency planning that proved immediately necessary the moment they landed on Norwegian soil.

After the erroneous parachute drop that added distance to the planned offset landing of Operation *GUNNERSIDE*, the members realized that they were not able to reach their objective with their heavy load of explosives. Based on Tronstad's guidance and priorities during training, Operation *GUNNERSIDE* made the decision to carry only the explosives needed for demolishing the high concentration installation.[58] This adjusted plan proved to be sufficient to neutralize the heavy water production, and, most likely, it saved the mission, considering the extra distance Operation *GUNNERSIDE* had to cover by skis.

In addition to the preparations that were enabled by the local knowledge, the reconnaissance mission, and the rehearsals; detailed preparations of equipment also proved to be important for the mission. US doctrine for special operations (Joint Pub 3-05) identifies "specially equipped" as a part of the definition of SOF.[59] This is often misinterpreted as a matter of high technology, but in practice the important issue is to have optimized equipment. SOB understood the need for optimizing *GUNNERSIDE* and *GROUSE* for the environment; consequently, both team leaders procured and designed their own special equipment. Based on their winter skills, the two team leaders requested clothing, skis, and rations that were not standard issue. The team leaders' efforts in these minor, but important, improvements demonstrated both the thoroughness of their preparations and their respect for the winter conditions in the area of Vemork.

Especially for Operation *GROUSE*, the optimized equipment proved important for their prolonged stay on the mountain plateau. Altogether, these preparations enabled the two patrols not only to carry out an audacious plan, but also to remain flexible and implement successful changes at all stages where friction challenged the original plan. Conclusion When the Allies assessed possibilities for destroying the German controlled heavy water production at Vemork in late 1942, the strong terrain and harsh winter conditions were the main challenges for a military operation. The British Combined Operations Command found itself limited in courses of action for the operation and the first attempt, the glider insertion of Operation *FRESHMAN*, resulted in the death of 41 British soldiers.

This conventional operation had limited flexibility, lacked contingencies, and failed in the rough weather and terrain of Norway. After the failure of Operation *FRESHMAN*, the two small teams for Operations *GUNNERSIDE* and *GROUSE* experienced success when they destroyed the heavy water production and escaped without either firing a single shot or causing collateral damage. The important key to success for Operations *GROUSE* and *GUNNERSIDE* was the missions' special operations nature.

The SOF personnel had skills that enabled them to remain flexible when faced with the friction of weather, terrain, and changing plans. The ability to adapt to these frictions helped them overcome major challenges to their plans and was vital for the success at Vemork.

As another example of the effectiveness of special operations over more conventional approaches, it is important to consider subsequent operations at Vemork. German efforts to reconstruct the heavy water plant in the nine months following Operation *GUNNERSIDE* led the Allies to launch a new conventional operation: the bombing of Vemork in November 1943. The result of the bombing proved true the challenges of the operating environment at Vemork as the underground heavy water installations remained intact. Additionally, the bombing resulted in collateral damage, including twenty Norwegian civilians killed at Rjukan.[60]

Again, SOE had more success in a following mission carried out by SOF personnel. The sinking of a Norwegian ferry carrying heavy water caused civilian casualties but achieved the objective of denying Hitler this strategic resource. Patrol sized operations achieved a strategic outcome with their efforts in February 1943, with a precise hit on the most vulnerable point of the heavy water production that SOE accomplished without losing personnel and without causing collateral damage. The small group of selected SOE soldiers capitalized on their specialized training, skills, and their unique motivation, when carrying out their mission in an extremely difficult operating environment.

The planning and execution of Operations *GUNNERSIDE* and *GROUSE* were textbook SOF operations that illustrate today' s theories of special operations. Like SOE during World War II, today's SOF units should carefully select, train, and employ only the limited number of soldiers that are able to fully master the operating environment. By mastering the operating environment and embracing the sources of friction rather than accepting their restrictions, SOF units can contribute to the current fight as flexible and creative elite warriors.

5. The Helberg interview (CIA Archive)

Studies Archive Indexes Vol. 36 No 3. The Vemork Action: A classic act of sabotage, Claus Helberg

> The following is an edited version of an article which was originally published in the 1947 Yearbook of the Norwegian Tourist Association. The author, who was a member of a Norwegian sabotage team, authorized its inclusion in *Studies in Intelligence* ... Mr. Helberg, who participated in the operation, In 1990 took a member of' the Editorial Board of Studies in Intelligence on a trip to Vemork, retracing part of the journey described in the article.

All mountain hikers still recollect that the Hardanger Plateau in Norway was blocked off by the Germans during World War II. From the spring of 1943 to the German capitulation, the plateau was a forbidden area, and the Wehrmacht confiscated the cabins in the mountains. Not even the cabins of the Norwegian Tourist Association were exempt. The Germans burned, plundered, and destroyed, and there were rumours circulating about clashes between British troops and Germans. The reason for the German actions gradually became public knowledge. It was because of the sabotage action in February 1943 against the heavy water production plant at Vemork.

As long as the war lasted, however, people were not given a factual explanation about these events, but they understood that something important was happening. The importance of the sabotage action was not disclosed until after the war against Japan was over. At that time, the War Department in London published the story on the Vemork action, and Churchill stated that the attacks against Vemork constituted a major event in the struggle for atomic power. The Germans had actually depended on heavy water from Vemork for their experiments with the atom bomb. As a member of one of the sabotage groups and as an employee of the Norwegian Tourist Association, I consider it appropriate to tell part of the story of what happened at Vemork and on the Hardanger Plateau during the war years.

How it Began In the summer of 1942, I was a member of the Linge Company's training camp in the mountains of Scotland. We were being trained for industrial and maritime sabotage and for training the Norwegian resistance forces. Professor (and Major) Leif Tronstad planned and organized the industrial sabotage. An excellent scientist, Tronstad was a member of the allied technical council. He made solid contributions to the Allies during the war, including the intelligence work behind the successful bomb attack against the German missile-testing site at Peenemunde. Tronstad was keenly interested in the heavy water production at Vemork. In 1934, as a chemical consultant, he had participated in planning the plant.

One day in August 1942, the training camp in Scotland was informed that four men, including myself, were to report to Professor Tronstad in London. In London, our team leader received instructions for the assignment. To keep the information on the planned attack from leaking, only he was briefed on the real assignment. The rest of us believed that the action concerned training the resistance. By the end of August, we had rounded up the skis, boots, and other winter equipment we needed. Then we had to wait for a moonlit night and fair weather. For the next six months, the weather was our worst enemy, not the Germans. On our first infiltration flight, one of the plane's engines burned up, and we limped back to England. The next time we had to turn back because of heavy fog.

Jumping In Finally, on 18 October, we had some luck. In the most beautiful moonlight, we circled over the Hardanger Plateau to find the target for our jump, the extensive Lokkje Marshes at Ugleflott. We believed that the navigator would have no trouble finding them. It is hard to describe my state of mind just before jumping. A few hours earlier, I had been soaking up sun on a lawn in England. The joy of returning home blended with the excitement of the jump itself. As the plane descended from 4,000 meters to 300 meters, the terrain no longer looked so fine and flat. "Action station!" The sharp command gave us something else to think about, and a few seconds later I was floating down to the plateau. We quickly learned that the pilots had not found Ugleflott after all.

We ended up in rough terrain, in boulders and rocky mountain slopes. But all four of us plus our 12 containers with food, equipment, weapons and ammunition landed nicely. Even our radio transmitter seemed to have survived the rocky landing. After a year in England, we were not quite used to the sharp, dry cold. Mission Revealed We soon got something else to think about, once our team leader told us about our actual mission. To assist in destroying the plant at Vemork, we were to find a good landing site for two British gliders with 40 Royal Engineers and to guide the men from the landing site down to Vemork.

We also were to reconnoitre the plant in advance and obtain all information about its German guards. The landing site could not be too far from Vemork. The British would not have skis, which meant that the site would have to be close to a road. The Skodal Marshes, 5 kilometres west of the Mos Lake Dam, were a logical choice. They are on both sides of the road between the dam and Rauland. Some Tough Going The operation was to be carried out during the next moon period, which was in one month. Unfortunately, we had landed 15 kilometres farther west than planned, at Fjarefit in the mountain range east of the Songa River. It is roughly 100 kilometres from Fjarefit to the Skoland Marshes. We had to pull 300 kilos of the equipment we needed, including food, skis, our radio, and weapons and ammunitions. This would have worked out well on good snow.

But in October we could not even expect snow for skiing. The snow that was on the ground melted the next day in the sun. However, we did not dare to go without taking our skis. If it should snow, we would be completely stuck without them. It took two days to collect our things, to pack what we would take, and to hide the rest in a depot. We had no time to spare. During the third night, heavy, wet snow began to fall. On 21 October we began marching to Mos Lake. For more than 14 days, we struggled in rough terrain and in indescribably bad snow. On good snow, an experienced skier would have covered that distance in a day or so. We just about exhausted ourselves to make it in 14 days. The Songa and Bitdal Rivers were not yet covered with ice, which meant that we had to struggle along steep birch-covered slopes.

Each of us carried more than 30 kilos, and that was only half of the equipment, so we had to backtrack and walk each distance twice. Under such conditions, we were tempted to leave behind as much as possible, either in the depot or along the route. We soon realized that we would have to resort to the strictest rationing to make the food last for four weeks. This was our daily menu: 1/4 plate of pemmican, one handful of oats, some crackers, and some chocolate. Because of the toil, our need for calories was much higher than normal. Each day, we became weaker and weaker.

Toward the last day we were not able to cover more than 2 kilometres. Fortunately, there were many cabins along the Songa and Bit Valleys, so we did not have to camp out. If we had had to sleep outdoors, I doubt we would have made it. On 10 November, we reached the Sandvass cabin, about 10 kilometres south of the Skodal Marshes, where we were to live until the glider party's arrival. On the trek from Fjarefit, our radio operator tried to contact England, but succeeded only once due to a defective battery. To get a new battery and to supplement our scanty food supply, I went down to the well-known skier, Torstein Skinnarland, who was the guard of the dam.

Through him, we got all we needed, and from then on the radio link worked smoothly. Skinnarland fed us information on Vemork, and after two of us had been down there to do a little spying, we were well prepared for the assignment. Ill-fated Attempt On 19 November we were told over the radio that the gliders would be coming the same night. Later in the evening, we went down to the landing site and put out our reception lights. But the wind increased, and it seemed that London had second thoughts.

We were informed over the radio that the operation was postponed for an hour. At 2300 hours, we heard the sound of planes. It certainly was a marvellous feat that the pilots had found the place in this weather. The lights were lit, and we expected every moment to see the gliders land. A little later, we heard plane engines, but now black cloud banks were drifting along the mountains, so I assumed that the chances were small for our lights to be seen. The next day, we received a sad report from England.

Only one of the planes had returned. One glider had been dropped over southern Norway, and the other plane with a glider was missing. We had nothing else to do than to wait for further orders. We did not feel very safe in our cabin. The German camp at the Mos Lake Dam was only 5 kilometres away. Consequently, we went northwest again, toward Saura, between Mogen and Kalhovd. There was a little hunting cabin there that was not marked on the map, and it became our home for three months.

More Hard Times This period offered us a fair share of difficulties. At the outset the Germans were especially worrisome. One day when I was down at the Mos Lake to get the latest news from our contact, Torstein, I was told that he had just been arrested. There had been large raids in Rauland, Mosstranda and Rjukan, and many people were arrested.

A German radio-direction-finding car had been observed at Rjukan, and German patrols had headed for the mountains. Every house in Rjukan and the Vestfjord Valley was searched, and the Germans asked everywhere if people had seen Englishmen in the area. The reason for the raids was obvious. It was a known fact that the British from one of the gliders had been captured and killed by the Germans. In the glider, the Germans had found maps and information showing that Vemork was their target. The Germans also almost certainly deduced that a radio transmitter in the area had been in contact with the British. Their suspicion fell on Einar Skinnarland, Torstein's brother. But because he was not at home when the Germans came to arrest him, they took Torstein instead.

The Germans were closer to their target than they realized. Einar belonged to our own group. He had, however, left England as early as April 1942 to furnish the Allies with information on German activities in the Rjukan region. We had not been in contact with Einar yet, but he joined us later in the winter. After we lost our contact with Torstein, our food situation became difficult again. For two weeks, we lived on oats, some margarine, sugar and Iceland moss. We dug the moss up from under the snow and boiled it. It hardly gave us any nourishment, but we imagined it did, and that helped a little.

On this diet, our strength waned quickly, and we were susceptible to diseases. At one time, we were all in bed, except the chief. I believe he managed to keep healthy because he spent much time thinking about reindeer steaks. He had shot many reindeer in these mountains before the war. The others in our group who had hardly seen a wild reindeer before were more sceptical about his chances of killing one. Meanwhile, the radio connection with England was maintained. We were told a new party would be sent, but only one plane would be used, and that it would deliver no more than six men, all from the Linge Company. The fellows would parachute to our quarters at Saura.

So, we would be 10 in all. The German guard force at Rjukan and Vemork numbered between 200 and 300 men. But we thought it would work out. Shortly before Christmas, there was a lot of rain and fog. Such troublesome weather suited us admirably. The Germans got tired of their patrolling and the wild reindeer came down from the western areas of the plateau. From then on, our chief was out daily to bag an animal, but conditions were not good for hunting. The fog was thick in the mountains, and he was so weak that he could only get to the nearest peak. But on Christmas Eve we received a welcome gift, clear and fine weather. Our patient leader went out again; this time he returned with reindeer meat. For two months, we ate only reindeer meat, as many as 11 animals.

To prevent scurvy, we ate the stomach and its contents from every animal. It tasted quite good, when mixed with blood and fat. Tough sledding. There was miserable weather in the mountains that winter, and we could not go hunting often. For weeks, the big storms would rage, and it was almost impossible for us to go outside. From time to time, we had a short break when the storms changed from east to west or the other way around, and then we had to make the utmost use of our time to go hunting or pick firewood. Fortunately, Saura was blessed with a birch forest. Polar disease is the name of a disease attacking people who live in isolation under difficult conditions. In a short time, the best friends may become the most bitter enemies. Even the most cheerful people may collapse. The fact that we did not fall victim to diseases was largely due to our leader's creative ability. We had no light, which meant long evenings.

Idleness is the cause of polar disease to a great extent. To prevent this, our chief started a study circle. Each evening, one of us gave a lecture. The theme did not matter; the main thing was our thoughts were removed as much as possible from our daily existence. From time to time, I went for a walk down to Rjukan or to the Mos Lake. We has a couple of contacts who furnished us with information on the movements of the Germans in the Vestfjord Valley. Meanwhile, the situation at Vemork had become less favourable.

It looked like the Germans were prepared for an attack, as the number of guards at the plant increased steadily. The Austrian guards were replaced by Germans; there were double guards where there had been single ones, and mines were deployed. Mounted floodlight facilities were installed so that the whole factory area could be lighted up as bright as day in case of alarm. We sent reports on these developments to England. The transmitter demanded much electricity, so we made strenuous trips with the battery to and from the charger. And I had to make many return trips.

Often, I was unable to force my way through the storms.

Contacting England. A New Group Arrives Fair weather finally returned one day at the end of February. It was real "drop weather," with moonlight and quiet, but 34 degrees below centigrade. Over the radio, we were told that the men would come that night, which meant that we would stand outside with the lights. We stood there in the cold until 0500, but nobody came. The next day we were told that the plane had not been able to find us and that it had dropped the men on the High Plateau, most likely 50 kilometres north of our position.

At the same time, we were hit by a terrible storm, the worst we had experienced. We had to stay in the cabin for a week. "*Gunnerside*" was the code name of the new group. They were given a rough reception: the big storm struck even before they had collected all their equipment. They were lucky, however, as there was a large cabin right by the landing site. On the first day of good weather, they headed south, and in the evening, we saw them coming from the west. It was fun to talk with friends who had news to tell. We had lived for five months alone in the mountains, and the change was welcome.

Attack Planning we immediately began to plan for the attack. We had to figure out how to get into the factory area and from there into the factory itself, and then how to get back alive. We had two alternatives to get to our target.

One was the bridge across the gorge by Vemork, but it was guarded. Liquidating the guard would make noise and perhaps reduce our small force. Another way was to try to cross the gorge under the bridge, and from there move unseen into the factory. We knew that the Germans believed that no one would try it that way. Our contact from Vemork had told us that no one could cross the gorge. But we doubted that. Local people often do not know their own area all that well, when you boil it right down. On the last day before the attack, I left the others to investigate on my own the possibility of getting across the river gorge. In the bright daylight I strolled down the country road past Vaer.

A little farther down, I left the road and headed for the gorge. I was able to make it down to the river ice, and I was also lucky enough to find a somewhat passable way up the factory side. It would not be as easy to go this way in the dark, but it was better than trying to get past the German guards on the bridge. On the basis of my information on the way across the gorge, our plan for our advance and attack was simple. As it got dark, we would start down the Fjosbu Valley, and then hike down into the Vestfjord Valley until we reached the country road. We would then follow the road about one kilometre to the spot where I had found the ford, move across the river and up on to the railroad line between Rjukan and Vemork on the south side of the valley.

We would follow the railroad up to the fence surrounding the factory area. After cutting our way through, we would split into a cover party and a blasting party. The cover party would take up positions so it could make a surprise attack in case the Germans sounded the alarm. If all stayed quiet, everyone was to remain at their posts until the explosion was heard. If the alarm was sounded during the advance, the cover party would attack the guards immediately. When the explosion was heard, they should assume that the blasting party was outside the factory area.

An order to withdraw would then be given. The passwords would be: "Piccadilly? Leicester Square!" The blasting party would destroy the heavy water plant in the basement of the Hydrogen Factory. They would try to find an open door into the factory. If they did not succeed, they would try to crawl through a tunnel for pipes and cables leading into the building. If something unforeseen happened, everyone was to act on his own initiative to complete the mission. Closing In It was overcast and there were heavy winds when we went down into the valley with skis and packs.

We all wore British uniforms. On the road, the ice was as slick and shiny as an ice rink. Off and on, the wind came in such strong gusts that we fell on the road. The weather could not have been better. We were the only ones out in such conditions. The climb down into the gorge went better than expected, but the river had risen because of the mild weather, and the water ran high above the river ice. But we got across and up to the railroad on the other side. When we were only 500 meters from our target, we stopped to go through the plan for the last time. It was vital that each man knew his assignment. We had a fine view of the bridge where the guards were posted. There would be a changing of guards at midnight, and because the guards' house was up by the factory building, we let the off-duty guards have 30 minutes to fall asleep.

Then we went up to the fence and cut our way through. I was with the cover party, and we took up our positions. We covered the German guardhouse and the entrances of the factory. Fortunately, we did not see a single guard. The noise from the factory was quite loud, and there was no other sound. It was pitch dark, and, even if a German had come, I do not believe he would have seen us leaning against the factory wall. The blasting party had some difficulties getting into the factory. The doors were locked, and it was not easy to find the cable tunnel in the dark. During the search, the men lost track of each other, and only two finally found the opening. They had the explosives, and the tunnel was just wide enough for them to crawl through.

They entered the room next to the heavy water plant. The entrance door was open, and they walked right in. There was a worker there, and he probably was scared stiff at the sight of the uniformed men. But he sensed the situation, and he kept quiet. Success The men started to place their explosives. They had practiced on wooden models in England and loads and fuses had been cut to size beforehand. While they were busy working, the silence was broken by the sound of glass being shattered nearby. One of the others in the blasting party who had not found the cable opening had decided to act on his own, and, when he found a window, he smashed it and crawled through.

He assisted in completing the work. To make sure that the explosives would not be removed, the fuses were cut as short as possible. The leader lighted the fuse, the worker dashed up the stairs to the upper floors, and the saboteurs ran out through one of the basement doors. The loads exploded 30 seconds later, and the heavy water instruments were destroyed. Except for smashed windows, only the instruments were demolished. The factory itself was undamaged. It may have been this fact that made the Germans react so slowly. One of the off-duty guards poked his head out the guardhouse door, turned a flashlight on the factory building, and then went back to bed! Thus, we got an unexpectedly large head start, and it was not until we were up on the country road that the first cars with Germans came up from Rjukan.

We found our skis and headed for the town.

Getting Away There was some discussion as to whether we should go straight up the mountain slope, but those of us who knew the area were able to veto that. The slope would have been too steep and heavy. Of course, it was risky to go down this road, because we could run into Germans at any moment, but we had no other way. The German cars went towards Vemork on the road right below us. Nobody must have thought that we would go toward Rjukan. At 0500, we were at Gvepseborg. And we felt great satisfaction when we sat and looked at Vemork and thought of all the commotion we had caused down there.

Then we headed for Hardanger Plateau again and for new assignments. Five men made straight for Sweden on skis, in English uniforms, with guns and ammunition. Our leader made his way alone to Oslo and then on to Sweden and England. Two went westward and established headquarters for the resistance forces in the Vinje region; one was back in England by Christmas, while the other stayed in Norway until the end of the war. The two telegraph operators established a radio station in the Hamre Mountains by the Mos Lake, and one operated this station until the war was over. A Hot Pursuit Many exciting stories can be told about each one of the men from that time on, but I will limit myself to my own adventures after our action.

First, I had to make a trip to Fjosbu Valley, and then I was to catch up with the others near Rjukan. But because of bad weather, I was delayed and did not find them. I therefore went alone to Geilo and took the train to Oslo. I left my uniform in the mountains. So far, the Germans had not started to search the Hardanger Plateau, and three weeks after the action I headed for the mountains again. I went from Imingen north to Skrykken. The purpose of my trip was to hide some weapons and explosives that had been left behind by *Gunnerside* at Skrykken. I found the cabin where the men had stayed and went in to get a little food after a long hike. The cabin looked like a pigsty, and it was obvious that the Germans had been there.

Perhaps they were close by! I ran out. Sure enough. Four or five men came running up towards the cabin. I put my skis on in a hurry and took off. My only weapon was a pistol. I had to rely on my abilities as a skier. After some shooting, one of the Germans was catching up with me. We were heading westward at great speed. I just wanted to have the evening sun in my face, to blind the Germans when they were shooting. But one German kept my speed. Fortunately, he also just had a pistol. Thus, the chances were equal, and the best shooter would win. The German evidently determined that the fight would be to my advantage.

He became so scared when I began to shoot at him that he emptied his whole magazine at me. All misses. Suddenly, our roles were changed. He threw himself around and headed back towards the other Germans. I hesitated for a moment, and then I pursued him as fast as I could. Every second was precious. His friends could show up over the crest at any moment. At a distance of about 30 meters, I fired a few shots at him. The sun had gone down, and darkness was approaching. I was safe, at least for the time being. But the ski tracks were a problem. The Germans could follow them, even in the dark. I knew there was no snow on the ice of Vra Lake, so I made for the lake. It was a starry night, but it was dark, and I had the bad luck of skiing off a precipice in the Slette Valley, just north of Vra Lake. I hurt my left arm badly, but I had to move on.

Down by Mogen, I barely avoided the Germans. They were swarming all over the place. It was obvious that they had started a major raid. But with a broken arm, I could not head for the mountains again. I had to see a doctor. Later in the day, I reached Hamaren, tired and hungry. I knew the people there, but I could not stay because the Germans searched every house at the Mos Beach. After I had had something to eat and rested a little. I moved across Mos Lake and down to Rauland. I had been on my feet for 36 hours and covered 160 kilometres. Rauland was standing on end. I could not have chosen a worse haven. There were about 300 German troops there. The weather turned mild, and it was impossible for me to consider moving on. I knew a merchant at Austbo, and I moved in with him and was allowed to sleep on his kitchen floor.

Germans had requisitioned the other rooms. Fast Thinking During the night, I concocted a story to tell the Germans, and they bought it! I said that I had reported as a local scout for the Germans in their search for the Vemork saboteurs and had broken my arm in their service. The Germans must have thought that I was quite a likable fellow. The German field doctor at Rauland examined my arm and sent me by ambulance down to Dalen, where I could travel on to Oslo by myself. In Dalen, I said *auf wiedersehen* to my helpers, and we parted as good friends. A boat was to leave the next day, and I wanted to rest a little. The tourist hotel looked inviting, and I found lodgings there. But bad luck struck again. Two German officers and their staffs came to the hotel in the evening and requisitioned most of the rooms. All the doors were blocked, and it was out of the question to try to escape.

All 30 guests were interrogated. We each had to give our name, address, age, and purpose of staying in Dalen. My identification card, which had been made in London, withstood a close examination. Under Arrest The next morning, 18 guests, including myself, were told to pack. We were arrested because of "insolent behaviour". I was in great doubt as to what to do, but I finally decided to go with the prisoner transport and then make a run for it. At any rate, I could not go all the way to Grini prison camp. I had put my pistol under the lining of my ski jacket. Before long, a big bus pulled up in front of the hotel, and we were accompanied to it, one by one. I made sure 1 was the last one to get out of the hotel, so that I would get a seat right by the exit door in the bus.

A German thought that I moved too slowly, and he gave me a kick in the rear so that I stumbled on the stairs. My pistol fell out of my ski jacket and landed between the feet of a German. Now I was in a jam! The Germans obviously did not know what to do with me, but the order was to send us to Grim, and an order is an order. One German guard was sitting in the front of the bus, and four others on two motorcycles convoyed us, one motorcycle 10 meters in front of and the other 10 meters behind the bus. My seat was on the floor in the rear of the bus. A young lady from Oslo sat next to me. We got to talking and actually had a cozy time. The German guard in the bus was envious, and eventually we changed seats.

I got to sit up front by the exit door, and he sat in the back. Of course, that made things easy. We had driven for nine hours, and it had become dark. In the Lier Hills between Drammen and Oslo, I opened the door and jumped out, amid the oaths of the Germans. I fell on the road but got back on my feet just as the Germans in the rear motorcycle were about to jump on me. Pursued by exploding hand grenades, I got into the forest, and, thanks to the darkness, my pursuers lost me. They gave up and had to go on without me. After benevolent treatment at Lier Asylum and a three-week stay at Drammen Hospital, I went on to England through Sweden. Another Mission In the night of 4 October 1944, nine soldiers parachuted at Ugleflott.

One of them was Major Leif Tronstad. My original team leader, another team leader, another team member, and myself were also among these men. Einar Skinnarland stood on the ground with the landing lights. Our assignment at this time was to assist the resistance force in Upper Telemark with weapons and training. It was of decisive importance for the economy of Norway after the war that the valuable industry in Telemark should remain intact. The Germans could be desperate enough to destroy the factories when they withdrew. The resistance would help to prevent this. Our headquarters was established in Skinn Valley by Mos Lake, with Tronstad as the chief.

During the winter, we requested and received more than 100 "drops." Almost every moonlit night, large Stirling planes came with guns, ammunition, food, uniforms and radio equipment. We had radio stations in every district, and we communicated among ourselves and with England. We established so-called cells, and the men in them were fully trained militarily. They were the elite troops. Training with weapons was given at night in the farm areas and in the towns. We had field manoeuvres in the mountains on Sundays. The Germans were more alert than ever, and their raids were part of our daily routine. We regarded the Hardanger Plateau almost as a recovered area. The blockade was still on, but the Germans had no chance of maintaining effective control there.

Then came 8 May. All the groups mobilized. On the mountain farms at the Mos Beach and Rauland, in every town and farm area, men and young boys dressed in uniforms. They had been prepared to defend their country. Fortunately, actions were not necessary. A long, liberating breath was let out by all Norwegian lungs. Professor Tronstad did not live to see that day.

On 11 March, he and another team member were shot by a traitor close to our headquarters. It was a great loss for Norway. Tronstad was an important scientist, and in no way could we afford to lose him. The people in the mountain farms around the Mos Beach mourned Tronstad deeply. Never had one person been so loved by the people on those isolated farms. All through the winter, he had lived with them. They appreciated his inexhaustible spirit and optimism, and they especially were fond of him for his considerate and friendly nature and for the warmth of his heart. For years to come, the memory of Professor Leif Tronstad will live on among the Mos Beach people.

ENDNOTES

Preface

1. Foot, M.R.D., *SOE – The Special Operations Executive 1940–1946* (London, BBC 1984), p. 211.
2. Kristoffersen, Major-General Frode, *Operations Gunnerside and Grouse – Special Operations during World War II against the German Controlled Heavy Water Plant in Norway* ('Master of Military Studies', Quantico, USMC Command Staff College, April 2010), p. 20. See appendix 4.

Prelude

1. This is the anglicized spelling, in Norwegian it is 'Hardangervidda'.
2. The Airspeed Horsa AS. 51 was Britain's initial response to successful German glider landings – the most spectacular being the attack on the great Belgian fortress complex of Eban-Emael in May 1940. The glider was named, perhaps appropriately, after the quasi-legendary fifth-century Saxon warrior.
3. The Handley-Page Halifax was a successful British design which entered service late in 1940 and was not retired for over two decades. A major element in Bomber Command's fleet, it was retained by many allied post- war air forces globally.
4. Combined Operations Headquarters was Churchill's response to the Allied defeat in France and the Dunkirk evacuation.

His idea was to combine land, sea and air forces to focus on planning amphibious operations against German-held coasts, working closely with the Joint Chiefs of Staff. Led initially by Admiral Roger Keyes, then Lord Louis Mountbatten and after 1943 by Major-General Robert ('Bob') Laycock.

5. A wartime RAF base located just northwest of Wick, in Caithness.
6. The Rebecca/Eureka transponding radar was a system of short-range radio navigation, specifically intended to facilitate dropping paratroops and their supplies. The Rebecca component was the airborne transceiver and antenna and Eureka was the ground-based transponder.

1 Scourge of the Swastika

1. Ohrstrom, Professor Lars, 'Leif Tronstad – the Heavy Water Hero', in *Chemistry World* (February 2018). See appendix 3.
2. Jomar Brun MBE (1904–1993), a native of Trondheim, graduated from the Norwegian Institute of Technology in 1926. Like Tronstad, he chose to resist and finally also had to flee to London. He survived the war and in 1951 took the chair in electrochemistry at the Technical Institute in his home town.
3. Ohrstrom, *op. cit.*
4. The project was a mighty ambitious one, Eyde, working with A/S Rjukanfoss required a budget roughly 2½ times that of Norway itself! This required significant outside investment and led to the company becoming Norsk Hydro.
5. Deuterium oxide (D2O), aka 'heavy water', is the form of water that contains two atoms of the ^{2}H, or D, isotope. The term heavy water is also used for water in which ^{2}H atoms replace only some of the ^{1}H atoms. In this case, rapid exchange between the two isotopes forms twice as many semiheavy HDO molecules as D_2O. The common hydrogen-1 isotope ^{1}H or H, also called *protium*) makes up most of the hydrogen in normal water. The presence of the heavier hydrogen isotope

gives this water different nuclear properties, and the increased mass confers different physical and chemical properties.

6. Vikdun Quisling (1887–1945) had come to some prominence through his work alongside the distinguished explorer Nansen and then orchestrated humanitarian relief efforts during the great Russian famine of 1921. Her served as ambassador there before returning in 1929 and subsequently served two terms as Minister of Defence. He set up his far-right party, the 'National Union' in 1933. The movement attracted members but not voters. In 1940, it was still very much a minority faction.
7. Ingolf Elster Christensen (1872–1943), Norwegian soldier and district governor. He was appointed head of the Administrative Council put up by the Supreme Court and for a while, the Germans toyed with the ideal of using him as head of state. Once the council was dissolved, he retired to his family farm and assumed no further role.
8. Membership slipped back and declined rapidly after it became increasingly obvious the Nazis were losing.
9. The Elverum Authorization.
10. *Freikorps* units began to appear in the wake of Germany's defeat in late 1918, a stew of ex-soldiers, paramilitaries, the disaffected and unemployed. They were the blunt tool used to wage confused wars in the east and suppress communists at home. Their numbers swelled to some 65 distinct corps.
11. *Reichswehr* (Reich Defence) were the regular armed forces of the Weimar Republic.
12. On 10 April 1940, a flotilla of five British destroyers under Commodore Warburton-Lee entered the Narvik Fjord and attacked German shipping with their naval escorts. The Germans under Kapitan Bonte outnumbered their attackers by more than 2 to 1 but were ill-prepared and suffered heavy losses to their merchant fleet. In the protracted fight, two British and two German destroyers were sunk and both commanding officers killed. Losses in personnel were roughly equal.
13. https://www.scallowaymuseum.org/,_accessed 29 January 2023.
14. Ibid.

2 *The Uranium Club*

1. In December 1903 Wells had published a short story in *Strand Magazine*, *The Land Ironclads*. A decade later, just before the outbreak of war, he wrote *The World Set Free* (London, McMillan, 1914), which predicted weapons of mass destruction, uncannily foreseeing the use of atomic weapons.
2. Tanks first saw action with dramatic if limited effect at Flers-Courcelette on the Somme that September.
3. Ramsay was a noted Scottish chemist who discovered the six noble gases (helium, neon, argon, krypton, xenon and radon) and won a Nobel Prize for Chemistry in 1904.
4. Ernest Rutherford, 1st Baron Rutherford, was a New Zealand physicist dubbed 'The Father of Nuclear Physics'.
5. Fritz Haber pioneered the use of lachrymose and toxic gases during the Great War. He won a Nobel Prize during the inter-war years before his fumigant substance Zyklon-B was used for deadly purposes during the Holocaust. Under a different branding, it is still available.
6. Goodman, M. S., 'MI6's Atomic Man: The Rise and Fall of Commander Eric Welsh' in *War in History* (Vol 23 (1) 100–114, 2016).
7. St Ermin's still exists as a luxury four-star hotel in Central London. It features Wi-Fi among its current attractions. I'm sure Welsh would have approved.
8. Paul Rosbaud (1896–1963), born in Graz, Austria, was a metallurgist and scientific advisor who worked for science publisher Springer Verlag (now Springer Science), which opened many doors in the scientific community. After the war he continued in publishing with Pergamon Press in Oxford.
9. 'MAUD' – Military Application of Uranium Detonation.
10. Sir Wallace Alan Akers (1888–1954) CBE FRS; after the war he returned to the ICI board. In retirement, he remained active in academe and research. He was one of the triumvirate whose deliberations helped found the UK Atomic Energy Authority.
11. Otto Robert Frisch (1904–1979) was an Austrian physicist who had worked with Lise Meitner on the theory of nuclear

fission. Teamed with Peierls, he designed the first theoretical trigger mechanism for an atomic bomb.

12. Sir Rudolf Ernst Peierls (1907–1995), was a German physicist who, with Frisch would play an important role in Tube Alloys and then in the Manhattan Project. His obituary in *Physics Today* hailed him as 'a major player in the drama of the eruption of nuclear physics into world affairs'. Such irony that Germany's religious intolerances denied the Reich such prodigious talent.
13. https://www.atomicarchive.com/resources/documents/beginnings/frisch-peierls-2.html :~:text=On%20the%20Construction%20of%20a,as%20a%20source%20of%20energy, accessed 09 February 2023.
14. Ecole Libre des Sciences Politiques.
15. The *Abwehr* (which translates as 'resistance/defence') was the intelligence arm of the *Reichswehr* and, after 1935, the *Wehrmacht*. Its relationship with the SD was not always cordial. Admiral Canaris had sacked Heydrich for ungentlemanly conduct, a humiliation that the status-hungry Reinhard, never forgot or forgave.
16. When France fell in June 1940, Joliot-Curie had already ensured the now French stock of heavy water (185 kilograms) was safely secreted across the Channel aboard a British vessel. All 26 flasks were concealed in the vaults below Windsor Castle until Tube Alloys technicians went to work on the magic substance.

3 The Ministry of Ungentlemanly Warfare

1. Cruikshank, C., *SOE in Scandinavia* (Oxford, OUP 1986), p. 55.
2. Errol Flynn did in fact star in a 1943 film, *Edge of Darkness* directed by Lewis Milestone, which featured just such a mass uprising wherein the vicious Nazi thugs in a small Norwegian port get their just deserts.
3. Cruikshank, op. cit., p. 56.
4. Ibid.

5. Sir Frank Nelson (1883–1966), was a career civil servant who, having served in the Indian Army during the previous war and then in the Indian Civil Service, returned to the UK and in 1924 was elected as MP for Stroud. When Germany invaded France, he was acting as consul in Basel but was successfully evacuated and then took an active role in the early stages of SOE. He retired due to ill-health in 1942.
6. Named after its Electra House premises and run by newspaper tycoon Sir Campbell Stuart.
7. Re-badged as MI (R) early in 1939.
8. Churchill's Secret Army, the Home Guard Auxiliers were also established by MI (R) during the same period.
9. Edward Hugh John Neale Dalton (1887–1962), Labour party economist and politician, helped shape Labour's foreign policy before the war and served as chancellor in Attlee's post-war cabinet.
10. The Massif de Vercors became a defiant redoubt for the maquis during July–August 1944, a doomed defiance ruthlessly crushed by the Germans – ill equipped and largely untrained resisters were no match for the *Wehrmacht*. See Paddy Ashdown's *The Cruel Victory* (London, William Collins) 2015.
11. Its creator was Lieutenant-Colonel F. W. Nicholls, a signals officer who had worked with Gubbins during the interwar years.
12. Leo Marks (1920–2001) MBE, was the son of a well-known antiquarian bookseller who, even at an early age, showed a marked aptitude for codebreaking. His chance came early in 1942 when he was conscripted, and he very quickly showed his mettle. Had he been less of a misfit or perceived misfit, he might have ended up at Bletchley, but he was effectively dumped on SOE. He proved to be a cryptographer of outstanding capacity and his influence was soon felt. He was responsible for training some of the most famous SOE agents.
13. Geoffrey Boothroyd (1925–2001) was a noted UK firearms expert and was widely published on the subject. He never had anything to do with SOE but contacted Fleming to point out that Bond's choice of pistol, a .25 Beretta, was largely

ineffective and subsequently advised on the agent's improved choice of handguns.

14. The Sten, amongst its other peccadilloes, tends to jam at embarrassing moments, most notoriously perhaps when Czech SOE agent Josef Gabcik stepped in front of Reinhard Heydrich's Mercedes on the morning of 17 May 1942. He had his target dead ahead, but the weapon jammed. It was Jan Kubis who threw a homemade grenade, inflicting injuries which eventually proved mortal.
15. William G. De Lisle invented the carbine that bore his name; a short-barrelled, fully suppressed (silenced) weapon that fired .45 subsonic ammunition, as near truly silent as any firearm is likely to get. Fewer than 130 were in fact manufactured.
16. The Welrod ('Wel' from Welwyn Garden City and 'Rod' from gangster slang), was a bolt action, six-round magazine, suppressed pistol in .32 ACP, developed by a major Hugh Reeves. A rather odd looking, ungainly weapon without the grace of, say, a Frontier Colt revolver, but over 14,000 were made and they were being deployed well into the Vietnam War era.
17. Fairbairn, W.E., *All-In Fighting* (London, Naval & Military Press 1942), p. 8.
18. Foot, M.R.D., *The Special Operations Executive 1940–1946* (London, BBC 1984), p. 61.
19. Ibid.
20. Ibid., p. 69.
21. Ibid., p. 70.
22. A total of 530 Norwegians would serve with *Kompani Linge*; of these 57 were killed.
23. Captain Graeme Black was posthumously awarded the DSO while Captain Joseph Houghton received an MC. Visitors to Sachsenhausen can still view the firing post if they so choose. Two Norwegian corporals from SOE were Erling Djupdraet (killed) and Sverre Granlund (escaped).
24. Herrington, I., 'The SIS and SOE in Norway 1940–1945: Conflict or Cooperation?' (in *War in History*, 2002 9 (1) 8–110).

25. *Galtesund* was built in Copenhagen in 1905, a coastal freighter which passed through several owners before being refitted for coastal ferrying from 1920. Seven years later, she began service on the Oslo–Kristiansand route as a passenger/cargo ferry. After her dramatic dash in March 1942, she served the allies as a tough little workhorse, only finally being withdrawn and broken up in Holland a decade after war's end.

4 *Grouse*

1. Helberg, CIA archive, op. cit.
2. Haukelid, K., *Skis Against the Atom* (North Dakota, North American Heritage Press 1989), p. 47.
3. Ibid. Haukelid tells us that Kjelstrup was wounded fighting during the invasion when a bullet struck him in the hip. Handily the plumber was packing a pair of pliers that deflected the shot, but fragments still lodged in the bone. He was hospitalised but still contrived to escape when the place was captured by the Germans.
4. Ibid., p. 49.
5. Gallagher, T., *Assault in Norway* (Connecticut, Lyons Press 2002), p. 19.
6. Quoted in Mears, op. cit., p. 45.
7. Ibid., p. 46.
8. Helberg, CIA Archive, op. cit.
9. Haukelid, op. cit., p. 49.
10. Helberg, quoted in Gallagher, op. cit., p. 22.
11. Poulsson, quoted in Gallagher, op cit., p. 27.
12. Ibid., p. 3c.
13. Axis security services, *Abwehr, Sicherheitsdienst, Ordnungpolitzei*, deployed the *Gurtelpeiler* portable radio direction finder ('RDF') to identify and locate clandestine radio transmitters, the spider to the fly. Allied radio operators always had to keep looking over their shoulders and be mindful of the hunters lurking in the wings.
14. Poulsson, quoted in Gallagher, op. cit., p. 31.
15. Haughland, quoted in Gallagher, op. cit., p32.

16. This is more than a little arbitrary, not to say subjective but I'd classify the Telemark story into five distinct episodes: 1. *Grouse*, 2. *Freshman*, 3. *Swallow*, 4. *Gunnerside* and 5. the bombing raid and SF *Hydro*.
17. Helberg, CIA Archive, op. cit.

5 *Freshman*

1. Wiggan, W., Operation *Freshman* (London, William Kimber 1986), p. 19.
2. Later Lieutenant General Sir Frederick Arthur Montague 'Boy' Browning (1896–1965), regarded by some as 'The father of British Airborne forces', a stylish old Etonian with a penchant for designing his own uniforms, married to the novelist Daphne du Maurier. His career rather stalled after the failure of *Market Garden* in September 1944. He had persisted in going ahead even though he had doubted the mission might be 'a bridge too far' and overlooking intelligence which suggested the enemy was far stronger and better prepared than expected. William Buckingham in his definitive *Arnhem: The Complete Story of Operation Market Garden 17-25 September 1944* pulls no punches in condemning Browning.
3. Brigadier Mark Chandos Auberon Henniker (1906–1991), commissioned into the RE 1926, served in India and then France 1939–1940, appointed commander RE 1st Airborne Division 1942. After the failure of *Freshman*, which could not be laid at his door, he went on to serve in the North African campaign and then in North-West Europe, serving after the war until his retirement in 1958. He was honorary colonel of the Parachute Engineers Regiment 1958–1968.
4. As part of Britain's awakening response to the successes of German airborne operations, the Glider Pilot Regiment was formed in December 1941. In essence the pilots were an elite within an elite, tasked to fly and navigate the cumbersome gliders but, once on the ground, to fight as infantry. They remain largely unsung, but their contribution was enormous and the risks terrifying. *Freshman* would be their first and

hardly encouraging foray. The regiment was disbanded in September 1957.

5. The village is in Rogaland County in southwest Norway and a memorial there commemorates the *Freshman* casualties.
6. These were led by Arne Lima who subsequently gave his captors the slip, making it to Sweden and then to Britain where he joined the Free Norwegian forces, Wiggan, op. cit., p. 60.
7. Quoted in Wiggan, op. cit., p. 81.
8. Major-General C.H. Miller was President of the court, members were Lieutenant-Colonel J.M. Ripley RA, Lieutenant-Colonel H.J. Wright RASC, Major the Honourable R.J. Palmer Coldstream Guards, Major H.J. Jennings RA with Major E.D. Wiltshire RAOC as waiting member; Major E. Steel, barrister-at-law Legal Staff Officer, Deputy-Judge Advocate's Staff, British Land Forces, Norway, sat as Judge Advocate. Major A. Burnett, RA Major Legal Staff, Military Department Judge Advocate's General Office, appeared for the prosecution; Captain L. Irving appeared for Seeling; Captain F.F. Millar for both Hoffman and Feuerlein. See Wiggan, op. cit., p. 110.
9. The doctor's defence was that he administered the morphine entirely as a palliative, believing the men were to be shot on the shore and this medication would ease the agonies, all being wounded, of their final journey. The tribunal remained unconvinced.
10. Seeling was executed on 10 January 1946, Hoffman on 15 May, Feuerlein was sentenced to life imprisonment but was subsequently handed over to the Russians by whom he had also been accused of atrocities against Soviet prisoners. He'd find his new captors rather less forgiving. The four unidentified British victims are now known to have been Corporal J.D. Cairncross (22), Lance-Corporal T.L. Masters (25), Driver P. Farrell (26), and Sapper E.J. Smith (24).
11. This time round, Major-General N.A. Coxwell Rogers was President of the Court with Colonel L.C.C Harrison; Lieutenant-Colonel J.C. de F. Sleeman, RTR; Major M.A. Gibbs of the Coldstreamers; Captain PD. Agnew, Irish Guards. The Judge

Advocate was Wing-Commander O.C. Barnet, barrister-at-law, appointed by the Judge Advocate General to the Forces. Captain G.G. Dromgoole, Judge Advocate General's Branch (War Crimes Section) HQ British Army of the Rhine led the prosecution with Dr R.G. Muller, a German lawyer appearing for the defence. See Wiggan, op. cit., p. 136.

12. Behrens was acquitted.
13. The trial began on 29 July 1946 with Major-General L.S.O. Playfair presiding. The members were Captain C.S.B Swinley, RN; Wing-Commander F.E. McDonnell, RAF; Major F.E. Gosling, 1st (The Royal) Dragoons; General-Major G. Meyer, Norwegian Armed Forces. Sitting - G.L. Stirling, Deputy-Judge Advocate General to the Forces. Prosecuting counsel: Colonel R.C. Halse and Major Bell-Macdonald, both Military Department, Judge Advocate General's Office. Dr R.G. Muller appeared for the defence. See Wiggan, op. cit., p. 152.
14. Von Falkenhorst was sentenced to death by firing squad though this was subsequently commuted to life imprisonment, which worked out as only seven years. He was released in 1953 on grounds of ill-health, though in fact he hung on until 1968.

6 *Swallow*

1. Poulsson, quoted in Gallagher, op. cit., p. 39.
2. Helberg, CIA Archives, op. cit. For the full version see appendix 5.
3. Ibid.
4. Ibid.
5. Wilson, quoted in Gallagher, op. cit., p.p., 40 – 41.
6. Ronneberg, quoted in Gallagher, op. cit., p. 41. Henry David Thoreau (1817–1862) was a US naturalist, essayist, poet and philosopher; Sigrid Gurie Haukelid (1911–1969) became a successful Hollywood actress during the pre- and post-war era.
7. First Aid Nursing Yeomanry.
8. Haukelid, op. cit., p. 76.

9. Ibid., p. 77.
10. Ibid. An inexpungable logic, worthy of Vladimir Putin today, though we may doubt the local landowners were fooled either.
11. Poulsson, quoted in Gallagher, p.p., 61 – 62.
12. Helberg, CIA Archives, op. cit.
13. Poulsson, quoted in Gallagher, op. cit., p. 65.
14. As a technical detail, the capsules were normally sewn into the collar of an agent's suit where it was easy to reach, more difficult in battledress with snow parka over the top, so the pills were shoved into the thigh pocket – next to the emergency field dressing!
15. Emily Dickinson; *A Thunderstorm* (1864).
16. Haukelid, op. cit., p. 87.
17. Ibid.
18. Ibid., p. 88.

7 Gunnerside

1. Haukelid, op. cit., p. 89.
2. Ibid., p. 96.
3. Ibid., p. 97.
4. Helberg, CIA Archive, op. cit.
5. Haukelid, op. cit., p. 97.
6. Helberg, CIA Archive, op. cit.
7. Ibid.
8. Haukelid, op. cit., po.p., 101 – 102.
9. Ibid., p. 102.
10. Helberg, CIA Archive, op. cit.
11. Haukelid, op. cit., p. 104.
12. Helberg, CIA Archive, op. cit.
13. Haukelid, op. cit., p. 105.
14. Ibid., p. 108.
15. Helberg, CIA Archive, op. cit.
16. Haukelid, op. cit., p. 108.
17. Helberg, CIA Archive, op. cit.
18. Haukelid, op. cit., p.p., 109 – 110.
19. Helberg, CIA Archive, op. cit.
20. Haukelid, op vit., p. 110.

21. Ibid., p. 112.
22. Helberg, CIA Archive, op. cit.

8 Escape!

1. Ronneberg, quoted in Gallagher, op. cit., p. 131.
2. Helberg, CIA Archive, op. cit.
3. Ibid.
4. Quoted in Mears, op. cit., p. 179.
5. Ronneberg, quoted in Gallagher, op. cit., p. 145.
6. Ibid.
7. Ibid., p. 147.
8. Helberg, CIA Archive, op. cit.
9. Ibid.
10. Ibid.
11. Ibid.
12. Ibid.
13. Ibid.
14. Ibid.
15. Ibid.

9 SF Hydro

1. This testimony of James Robert Ward is recorded as part of the Westall Project; http://www.northshields173.org, accessed 08 May 2023.
2. The Boeing B-17 was a fast four-engined strategic bomber which could carry a heavy payload a very long way. Massively armed and extremely robust, it was a major asset in the Allies' inventory of bomber aircraft.
3. The Consolidated B-24 came into service in 1941, like the B-17 it had four powerful engines and could heft a substantial payload, though its overall handling and performance were often compared unfavourably.
4. US Eighth Air Force was initially formed at Langley, Virginia in February 1942. Within two years it had matured into a vast warfighting machine with some 200,000 personnel and able to fly 1,000 aircraft in a single sortie. As with Bomber

Command, its successes came at a fearful price, with some 26,000 casualties overall.

5. https://www.b24.net/MM111643.htm, accessed 16 May 2023.
6. These losses were due to mechanical failure rather than enemy action.
7. I actually write these lines on the 80th anniversary of the Dambusters Raid, one of the great wartime legends of my youth and one which sold an awful lot of Airfix Lancaster bomber kits.
8. Mears, op. cit., p. 204.
9. SOE's Norwegian operations, whilst serious box office, didn't necessarily sit too well with Milorg who feared that blowing things up, generally stirring the hornet's nest, might read well in the Allied press but it risked compromising their equally valuable if longer term objectives of gathering intel and building up potential home forces.
10. Quoted in Mears, op. cit., p. 211.
11. Astley, J.B., *The Inner Circle* (Toronto, Little Brown 1971), p. 38.
12. Haukelid, op. cit., p.p. 182–183.
13. Ibid., p. 183.
14. Ibid., p. 184.
15. Ibid., p.p. 185–186.
16. Ibid., p. 186.
17. Ibid., p.p. 191–192.
18. Ibid., p. 192.
19. Ibid., p. 192.
20. Ibid., p. 193.
21. Ibid., p. 195.
22. The port of Herøya forms a peninsula in the municipality of Porsgrunn, situated between the fjords of Frierfjord to the west and Gunneklevfjord to the east, at the mouth of Telemarksvassdrageyt.
23. Haukelid, op. cit., p. 197.
24. Ibid., p. 198.
25. Haukelid's own assessment and that of the Oslo newspaper was that fourteen Norwegians and twelve Germans had

gone down with the ferry and lost their lives. Sorensen later put the death toll among the Axis guards at twenty-plus. An intelligence report, unearthed from Von Falkenhorst's files, claims that the full complement of passengers and crew was 53, of whom 27, including four *Wehrmacht*, were rescued.

10 Norwegian Spring Sunshine

1. Helberg, CIA Archives, op. cit.
2. These awards may be summarised as follows: For Leif Tronstad an OBE; For Poulsson, Haughland and Ronneberg each a DSO; MCs for Haughland, Idland and Haukelid and an MM for Helberg, Kjelstrup, Kayser, Storhaug and Stromsheim.
3. https://www.ronneberg.org/Stories/joachim.html, accessed 26 April 2023.
4. Ibid.
5. Ibid.
6. Gunnar Fridjof Thurmann Sonsteby (1918–2012), known as *Kjakan* ('The Chin') or just Agent no. 24, was the most highly decorated Norwegian resistance fighter of the war.
7. https://www.theglobeandmail.com/incoming/a-real-life-hero-of-telemark/article25278635/, accessed 05 May 2023.
8. Ibid.
9. Ibid.
10. Smith, General Sir Rupert, *The Utility of Force* (London, Allen Lane 2005), p. 197.
11. Many years ago the author attended a lecture given by the late Brigadier Peter Young, a much-decorated former commander of No 3 Commando (and founder of the Sealed Knot Civil War re-enactment society). At the end one questioner asked, 'Who was the most formidable enemy you faced during your war years?' He replied without hesitation, 'Unless you've fought the Germans, you've never been to war ...'
12. Kristoffersen, op cit., pp 4–5. For a fuller version, see appendix 4.

Appendix 4

1. Heavy water contains a higher proportion than normal water of the isotope deuterium. German scientists needed heavy water in large quantities as they used it as braking fluid in the chain reaction within the nuclear reactor. Jens-Anton Poulsson, *The Heavy Water Raid: The Race for the Atom Bomb 1942-1944* (Orion Forlag AS, 2009), 89.
2. SOE report, 'S.O.E. and "Heavy Water"', London, April 1944, 1.
3. Robert G Spulak: Jr; A Theory of Special Operations, JSOU report 07-7 (The JSOU Press, Hurlburt Field, Florida), 21.
4. William H McRaven, *SPEC OPS; Case Studies in Special Operations Warfare: Theory and Practice* (Presidio Press, 1996), 11.
5. McRaven, 11.
6. McRaven, 11.
7. Spulak, 21
8. Carl Von Clausewitz, *On War* (Everyman's Library, 1993), 140.
9. Clausewitz, 138.
10. Spulak, 21.
11. Spulak, 23.
12. Thomas Gallagher, *Assault in Norway. Sabotaging the Nazi Nuclear Program* (The Lyons Press, 2002), 17.
13. Ray Mears, *The Real Heroes of Telemark* (Hodder and Stoughton, 2003), 21. According to the author, the six different courses of action were: "1) Using men already employed at Vemork to blow up the stocks; 2) deploying agents to infiltrate the plant; 3) sending in an SOE sabotage party; 4) a Combined Operations attack using hydroplanes for arrival and departure; 5) sending in … a 'suicide squad', i.e. one that had no chance of escaping afterwards; 6) RAP bombing."
14. Gallagher, 20: The Norwegian scientist who served as an advisor for SOE pointed out that Rjukan, the electrical power centre of southern Norway, had large liquid-ammonia stocks, a flammable gas stored under high pressure in tanks.

15. Haukelid, 50.
16. Poulsson, 107.
17. SOE renamed Operation *GROUSE* in the new plans, but the members of the reconnaissance party never used the new name, Operation SWALLOW.
18. Haukelid. 6. (Introduction by General Major Sir Collin Gubbins, CO of Special Operation Executive.)
19. Haukelid, 78-79. Team *Gunnerside* waited three months for the right conditions for an insertion. In January 1943 the team came close but had to abort an attempt when they were over the drop zone.
20. Some sources (e.g., Haukelid, 74) include Einar Skinnarland and count 11 members in the heavy water raid. Skinnarland was a SOE member and parachuted into Rjukan six months prior to Operation *GROUSE*. He worked as a covert agent and was deeply involved in the heavy water raid, but not a part of either *GUNNERSIDE* or *GROUSE*.
21. Joachim Ronneberg, Operation *Gunnerside*. Fenrik Ronneberg's Report. London, April 1943, 15. Half of the group had received follow-on missions in Norway and did not exfiltrate to Sweden.
22 Ray Mears, *The Real Heroes of Telemark*, 22. 24
23. Knut Haukelid, *Skis Against the Atom* (North American Heritage Press, 1989), 2. (Introduction by General Major Sir Collin Gubbins, CO of Special Operation Executive.)
24. Spuıak, p 1. "Special Operations Forces (SOF) are small, specially organized units manned by carefully selected people using modified equipment and trained in unconventional applications of tactics against strategic and operational objectives."
25. Mears, 31.
26. Pouısson, 90. Poulsson explains how his team member Haugland was given priority on the flight from Sweden because of his skill as a telegraph operator. Sverre Steen, *Norges krig 1940-1945* (Gyldendal Norsk Forlag, 1947), 663. It describes the flight, the *Stockholmsruten* from Sweden to Scotand, and its limitations.

27. Pou1sson, 48.
28. Spu1ak, 12
29. Tomas Adam, email message to author, January 10, 2010. The four stages of training were: stage 1) basic training; 2) sabotage operations; 3) parachute training; 4) mission-specific training.
30. Haukelid, 71.
31. Jostein Berglyd, *Operation Freshman, The Hunt for Hitler's Heavy Water* (Stockholm: Leandoer & Ekholm Forlag, 2006), 26.
32. Jens-Anton Pou1sson, interview with author, June 10, 2009.
33. Gallagher, 18. "When Combined Operations turned to SOE for advice, the Norwegian Section objected to the plan on the grounds that it was ill-conceived and susceptible to too many failures at too many stages in its operation." Gallagher continues with a list of nine hazards pointed out by Wilson and Tronstad of SOE. All nine are possible points of friction to the operation due to terrain or weather.
34. "Lieutenant Colonel Mark Henniker ... led the training and planned the operation ... He argued that the distance from the central base of the operations ... to the landing site in Hardangervidda was too long."
35. Haukelid, 79.
36. Poulsson, 108.
37. Berglyd, 30.
38. Ronneberg, interview with Tomas Adam, January 10, 2010.
39. Haukelid, 1 00.
40. Mears, 34.
41. John Berg, *Soldaten som ikke ville gi seg: Lingekaren Arne Kjelstrup 1940-45*, (Metope, 1986), 102.
42. McRaven, 11.
43. Berg, 146.
44. In a 10 June 2009 interview with author, the *GROUSE* team leader stated that in the conditions Operation *GROUSE* faced, no one cou1d have second-guessed a decision to abort the mission.
45. Pou1sson, 104.
46. Spulak, 1.

47. Spulak, 20.
48. Headquarters U.S. Marine Corps, Marine Corps Planning Process, MCWP 5-1, (Washington, DC: U.S. Marine Corps, January 5, 2000), 2-3 and G-96: commander's 25 intent includes purpose, method, and end state of enemy and own forces (location and status).
49. Spulak, 20
50. Poulsson, interview with author, June 10, 2009.
51. Spulak, 20.
52. Poulsson, interview with author, June 10, 2009.
53. Poulsson, 133.
54. McRaven, 16.
55. McRaven, 10.
56. Haukelid, 71.
57. Lieutenant-Colonel Jamie Hammond, "Special Operations Forces: Relevant, Ready and Precise," *Canadian Military Journal*, July 14, 2008, http://www.journal.dnd.ca/vo5/no3/special-O 1-eng.asp (accessed February 10, 2010).
58. Haukelid, 86.
59. U.S. Department of Defense. Doctrine for Joint Special Operations, Joint Pub 3-05, April 17, 1998, GL 10.
60. Poulsson, 168.

SELECT BIBLIOGRAPHY

These events have been quite extensively chronicled since 1945. In May of that year, after interviewing agents Nielsen and Helberg, Barbara Wace of the Associated Press filed a news article on the sabotage. Wartime censorship delayed publication until after the atomic bombings of Hiroshima and Nagasaki in August 1945. Drummond, J. D., *But for These Men* (1962), recounts the raids on the Norsk Hydro heavy-water factory at Vemork and another the railway ferry, *Hydro*. Mears, R., *The Real Heroes of Telemark: The True Story of the Secret Mission to Stop Hitler's Atomic Bomb* (2003) stresses the Norwegian Commandos' unique survival skills. His book is a companion volume to *The Real Heroes of Telemark*, a BBC television documentary series.

Haukelid's *Skis Against the Atom* is a first-hand account by one of the *Gunnerside* raiders who stayed behind. Jens-Anton Poulsson wrote *The Heavy Water Raid: The Race for the Atom Bomb 1942–1944* (2009). Operation *Freshman* is covered in two books: Wiggan, R., *Operation Freshman: The Rjukan Heavy Water Raid 1942* (1986) and Berglyd, J., *Operation Freshman: The Actions and the Aftermath* (2009).

Richard Rhodes's Pulitzer Prize-winning book, *The Making of the Atomic Bomb*, includes details of the events, as does Leo Marks' *Between Silk and Cyanide: A Codemaker's Story 1941–1945*

(1998). Marks, SOE's cryptographer, knew the Norwegian team, trained them in cryptography so they could communicate with SOE in England, and followed their progress after they were dropped in Norway. The raid is the subject of interviews with many of the Commandos.

Operation *Gunnerside* is covered by Bascomb, N., *The Winter Fortress: The Epic Mission to Sabotage Hitler's Atomic Bomb* (2016) and in Damien Lewis's book, *Hunting Hitler's Nukes: The Secret Race to Stop the Nazi Bomb* (2016). *Heroes of Telemark; Sabotaging Hitler's Atomic Bomb, Norway 1942–44* by David Greentree (2018) describes the operation's planning, execution and aftermath.

Primary, Unpublished and Journal Sources

Central Intelligence Agency of the USA 'CIA Studies Archive Indexes', vol. 36, no 3, *The Vemork Papers*, Helberg article 1990, see: https://www.cia.gov/static/ccf14df21af9e69433e3341d281fa52c/The-Vemork-Action.pdf

Kristoffersen, Major F., 'Operations *Gunnerside* & *Grouse*; special Operations during World War II against the German controlled Heavy Water Plant in Norway' (April 2010)

Mann, C., 'British Policy and Strategy Towards Norway 1941 – 1945' (2012) https://dokumen.pub/british-policy-and-strategy-towards-norway-194145-978-1-349-30286-4-978-1-137-28435-8.html

National Archives HS/2/172, HS/2/173, HS/2/184, HS/2/185, HS/2/186, HS/2/187, HS/2/188, HS/2/189, HS/2/190

Norwegian Industrial Workers Museum, Vemork: NIA/A-1148 - Jomar Bruns arkiv

Ohrstrom, L., 'Leif Tronstad: The Heavy Water Hero' (*Chemistry World* 7 February 2018)

Resistance Museum, Oslo:

0012 SOE Dbbb - 0022 SOE 22 1942-1944 SA 004 2/3

15/2/1. Bullfinch. Rosy Telegr. okt. 43-mai 44
15/2/2. Bullfinch. Rosy Red Telegr. juli-Okt.44

15/3/1. Bullfinch Rosy. Rapporter
15/6/1. D 15. Venner (Friends)
15/8/1. D 15. Våpenlager
16/1/1. Grouse. Operation instructions
16/1/1. Swallow. Operation instructions
16/2/1. Hawfinch. Red. Telegrammer
16/2/2. Bullfinch Trilby Telegrammer
16/2/3. Grouse/Swallow Telegr. nov. 42-mars 43
16/2/3. Swallow Telegr. mars-juni 43
16/2/3. Swallow Telegr. juli-aug. 43
16/2/3. Swallow Telegr. Sept. 43-febr.44

0013 SOE Dbbb 0023 - SOE 23 1943-1944 SA 004 2/3

16/2/3. Swallow Blue. Telegr. febr-mai 44
16/2/4. Swallow Brown. Telegr. apr-aug. 44
16/2/4. Swallow Brown. Telegr. Sept-okt.44
16/2/5. Swallow/Primus Green Telegr. Sept.43-sept.44
16/2/5. Swallow Yellow Telegr. mai-okt. 44
16/2/6. Moonlight Telegr.
16/2/7. Lamplight Telegr.
16/2/8. Starlight. Telegr.
16/3/1. Swallow. Rapporter
16/3/2. Gunnerside. Rapport
16/3/3. Hawfinch Red. Rapport
16/3/3. Lamplight. Rapport

0014 SOE Dbbb - 0023b SOE 23b 1942-1944 SA 004 2/3

16/3. D 16. Diverse rapporter
16/4/1. Swallow. Telegr. og Rapporter
16/4/2. Sunshine. Telegr. og Rapporter
16/5/1. D 16 som operasjonsområde
16/6/1. D 16. Venner (Friends)
16/7/1. D 16. Foes
16/8/1. D 16.Våpenlager
17/3. Bonzo. Diverse rapporter

17/3. D 17. Diverse rapporter
17/3. Swallow/*Grouse*. Historisk rapport
17/5/1. D 17. Oppgave over forsyninger
17/6/1. D 17. Venner (Friends)
17/7/1. D 17. Foes
17/8/1. D 17. Våpenlager
Royal Engineers Archive, Royal Engineers Museum Chatham; AE/2/10, AE/2/19, AE/4/10
SOE and 'Heavy Water' (London SOE 1944)

Secondary Sources

Bailey, R., *Forgotten Voices of the Secret War* (London, Ebury Press 2009)
Bascomb, N., *The Winter Fortress: The Epic Mission to Sabotage Hitler's Atomic Bomb* (London, Head of Zeus 2016)
Berglyd, J., *Operation Freshman: The Actions and the Aftermath* (Solna: Leandoer & Ekholm 2009)
Cooper, D.F., 'Operation Freshman' in *The Royal Engineers Journal* 1945
Cruickshank, C., *SOE in Scandinavia* (Oxford, OUP 1986)
Drummond, J. D., *But for these Men* (London, W.H. Allen 1962)
Foot, M.R.D., *The Special Operations Executive 1940–1946* (London, BBC 1984)
Gallagher, T., *Assault in Norway* (Guilford, The Lyons Press CT 2010)
Greentree, D., in *Heroes of Telemark; Sabotaging Hitler's Atomic Bomb, Norway 1942–44* (Oxford, Osprey 1918)
Haukelid, K., *Skis Against the Atom* (North Dakota, North American Heritage Press 1989)
Koht, H., *Norway: Neutral and Invaded* (London, Hutchinson & Co., 1941)
Lewis, D., *Hunting Hitler's Nukes: The Secret Race to Stop the Nazi Bomb* (Quercus, London 2016)
Lloyd, A., *The Gliders* (London, Leo Cooper/Secker & Warburg 1982)
Marks, L., *Between Silk and Cyanide: A Codemaker's Story 1941–1945* (Stroud, History Press 1998)

Mears, R., *The Real Heroes of Telemark: The True Story of the Secret Mission to Stop Hitler's Atomic Bomb* (London, Hodder & Stoughton 2003)

Milton, G., *The Ministry of Ungentlemanly Warfare* (London, John Murray 2016)

Mountfield D., *The Partisans* (London, Hamlyn 1979)

Petrow, R., *The Bitter Years, The Invasion and Occupation of Denmark & Norway, April 1940–May 1945* (London, Hodder & Stoughton 1978)

Owen, J., *Commando – Winning WW2 Behind Enemy Lines* (London, Little Brown 2012)

Poulsson, J. A., *The Heavy Water Raid: The Race for the Atom Bomb 1942–1944* (London, Orion 2009)

Rhodes, R., *The Making of the Atomic Bomb* (London, Simon & Schuster 2012)

Smith, General Sir Rupert, *The Utility of Force* (London, Allen Lane 2005)

Warwicker, J., *Churchill's Underground Army* (London, Frontline 2008)

West, N., *MI6 Britain's Secret Intelligence Service Operations 1909–1945* (London, Weidenfeld & Nicolson 1983)

Wiggan, R., *Operation Freshman: The Rjukan Heavy Water Raid 1942* (London, Harper Collins 1986)

INDEX